Macmillan Revised ENCYCLOPEDIA of SCIENCE

5 Plants and Animals

Their Diversity and How They Live

John Stidworthy

Macmillan Reference USA
New York

Published by:
Macmillan Library Reference USA
A Division of Macmillan Publishing USA
1633 Broadway, New York, NY 10019

Copyright © 1991, 1997 Andromeda Oxford Limited
An Andromeda Book
Devised and produced by Andromeda Oxford Ltd,
11-15 The Vineyard, Abingdon, Oxfordshire, OX14 3PX England

Macmillan edition copyright © 1991, 1997
Macmillan Reference USA

Library of Congress Cataloging-in-Publication Data

Macmillan encyclopedia of science. -- Rev. ed.
    p.   cm.
    Includes bibliographical references and index.
    Summary: An encyclopedia of science and technology, covering
such areas as the Earth, astronomy, plants and animals, medicine,
the environment, manufacturing, communication, and transportation.
  ISBN 0-02-864556-1 (set)
  1. Science--Encyclopedias, Juvenile. 2. Engineering-
-Encyclopedias, Juvenile. 3. Technology--Encyclopedias, Juvenile.
[1. Science--Encyclopedias. 2. Technology--Encyclopedias.]
I. Title: Encyclopedia of science.
0121 M27 1997
500--DC20   96-36597
                                                                  CIP
                                                                   AC

Volumes of the *Macmillan Encyclopedia of Science*
Set ISBN 0-02-8645561
  1 *Matter and Energy*   ISBN 002864557X
  2 *The Heavens*   ISBN 0028645588
  3 *The Earth*   ISBN 0028645596
  4 *Life on Earth*   ISBN 002864560X
  5 *Plants and Animals*   ISBN 0028645618
  6 *Body and Health*   ISBN 0028645626
  7 *The Environment*   ISBN 0028645634
  8 *Industry*   ISBN 0028645642
  9 *Fuel and Power*   ISBN 0028645650
10 *Transportation*   ISBN 0028645669
11 *Communication*   ISBN 0028645677
12 *Tools and Tomorrow*   ISBN 0028645685

Printed in the United States of America

Introduction

We share the Earth with an enormous number of different kinds of living things. This volume of the encyclopedia surveys this amazing diversity and examines the varied life-styles and behavior shown by animals. (Additional information on plants and animals is in other volumes – evolution and basic life processes are discussed in Volume 4, and ecology and agriculture in Volume 7.)

To learn about a specific topic, start by consulting the Index at the end of the book. You can find all the references in the encyclopedia to the topic by turning to the final Index, covering all 12 volumes, located in Volume 12.

If you come across an unfamiliar word while using this book, the Glossary may be of help. A list of key abbreviations can be found on page 87. If you want to learn more about the subjects covered in the book, look at the Further Reading section.

Scientists tend to express measurements in units belonging to the "International System," which incorporates metric units. This encyclopedia accordingly uses metric units (with American equivalents also given in the main text). More information on units of measurement is on page 86.

Contents

Part One

The variety of living things

Almost everywhere you go on the face of the Earth you will find living things. From the ocean depths to high in the mountains there is life. Earth is full of living things. We do not know yet whether life exists elsewhere in the Universe.

Some of the forms of life are very big. An old oak tree or a whale dwarfs us. But the great majority of plants and animals are small. The most numerous, such as bacteria, can be seen by us only with the aid of microscopes. Well over a million different kinds, or species, of living things have already been named and described. This first part of the book introduces the main groups of animals and plants.

The variety of living things, and their many different ways of life, is amazing. Probably few large animals or plants wait to be discovered. But in tropical forests new species of small animals, such as beetles, can still be found.

◀ Bright sunflowers growing in a field in North Dakota. Sunflowers are widely grown for the nutritious oil in their seeds. The large, yellow flowers are attractive to bees, which pollinate them.

Lowly life

For at least three billion years there have been living things on the Earth. The earliest living things that we know of, bacteria and blue-green algae, are still common and important today. Bacteria and blue-green algae are tiny, each consisting of just a single cell. We can only see what they are like using microscopes. The simplest animals are the protozoa. These also consist of just a single cell. Most living things are made up of many cells. Among the algae we can see both single-celled and many-celled forms. More complex animals and plants have developed as time has gone on.

SPOT FACTS

• The smallest viruses are icosahedrons (20-sided polygons) that measure about 18 to 20 millionths of a millimeter wide.

• Viruses represent the last great challenge to medical science. There is no treatment for most viral illnesses, including the common cold and influenza.

• Just 1 g (0.03 oz.) of soil can contain up to 2.5 billion bacteria.

• In one day a single bacterium could multiply to become 28 trillion.

• The giant kelp seaweed can grow 60 m (200 ft.) long – and it lives for only a year.

• Some fungi make new threads up to 1 km (over 0.5 mi.) long in one day.

• "Red tides" occur when a sea is colored by vast clouds of tiny red algae. Some contain powerful nerve poisons. Some are so strong that 1 g (0.03 oz.) would kill 5 million mice.

• A bacterium may divide as often as every 6 minutes, very quickly forming a colony that is visible to the human eye.

• In the food chain, bacteria are decomposers. Without them, dead plants and animals would rapidly accumulate on the Earth's surface.

• The common field mushroom may form 12 billion or more spores on its fruiting body.

• Fungi help us in the fermentation of wine and in the manufacture of citric acid and antibiotics.

• Algae were discovered in 1984 on a Bahamas seamount 268 m (879 ft.) down, a great depth for a light-dependent plant.

• The highly absorbent *Sphagnum* moss was used in the Middle Ages to line babies' diapers.

• Lichens are a "combination plant" of fungus and algae, and are useful sources of colorants. Archil, a purple dye made from lichen, is used as a food colorant and to form litmus, the acid-alkali indicator.

• Reindeer moss, important as the winter food of moose and caribou, is actually a lichen. It readily concentrates pollutants and radioactive materials, which can cause harm to animals that eat it.

• The desert lichen can be blown around because it is only loosely attached, and may be the manna from heaven of the Bible, the food that miraculously appeared to the Israelites in the wilderness.

• During the Lower Carboniferous and Upper Carboniferous (Mississippian and Pennsylvanian) periods (about 300 million years ago), ferns were the dominant form of vegetation.

• Many epiphytes, or air plants, are ferns. They attach themselves to trees and live on airborne moisture and dust particles.

• Horsetails are related to ferns. They often contain glasslike silica and, when dried, can be used for polishing wood and cleaning metals.

Microorganisms

Bacteria and blue-green algae have a cell wall but no nucleus or specialized parts within the cell. Bacteria live almost everywhere – in soil, air, ocean depths, even in hot springs at 92°C (198°F). Some use sunlight to make food as plants do. Others make energy from other chemical reactions, such as turning sulfur to sulfuric acid. Many are useful to us and other living things. Bacteria trap atmospheric nitrogen and turn it into substances that act as fertilizers for plants. Bacteria also cause many diseases, such as cholera and tetanus.

Single-celled animals have a nucleus and a soft cell membrane. Protozoa such as ameba live in fresh water; some live in the sea and soil. Some are parasites that cause diseases such as malaria and sleeping sickness.

Viruses cannot survive by themselves, but are found inside the cells of other living things. There they make copies of themselves. They cause a variety of diseases.

▼ The sharp point of a pin (1) is magnified 175 times by a scanning electron microscope. The tiny rodlike shapes clustered in the lower part are bacteria. At a magnification of 3,535 times is a microscope view (2) of *Lactobacillus bulgaricus* bacteria in yogurt. These bacteria, visible here in the pink tubes, make live yogurt a healthy food to eat. The harmful bacterium *Salmonella typhimurium* is shown in (3). *Salmonella typhimurium* is transmitted in contaminated food, and can cause food poisoning in humans, with symptoms of diarrhea and fever. This electron-micrograph picture shows the effect of the antibiotic drug *Polymyxin B* on this bacterium.

3

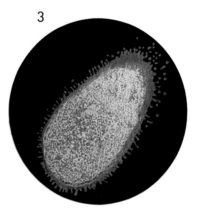

2

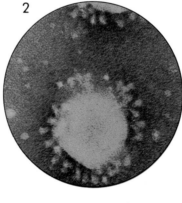

1

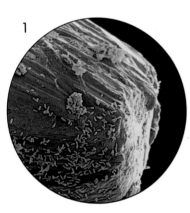

▼ Some fungi can be seen with the naked eye. Smaller fungi, and the protozoa, need a light microscope to be seen well. Bacteria may be visible in a light microscope, but details need the greater magnification provided by an electron microscope. Viruses are so small that they can be seen only with the help of an electron microscope.

Magnifications

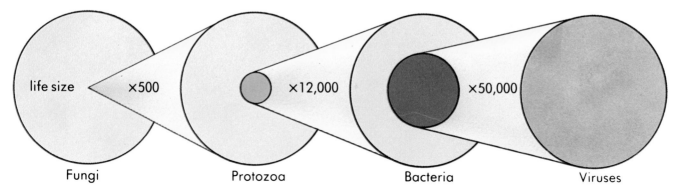

life size ×500 ×12,000 ×50,000

Fungi   Protozoa   Bacteria   Viruses

Algae

Algae are simple plants that do not have true stems, leaves, or roots. Many of them are single-celled. These include the diatoms, which have a shell like a carved pillbox. They live drifting in the sea or fresh water, and are food for many animals. Dinoflagellates also have shells around their single cells. They live in the sea, and have two flagella. These threads stick out from the cell wall and beat, spinning the dinoflagellate as it moves through the water. *Euglena* and its relatives live in fresh water. *Euglena* has a single flagellum.

Euglena is green, and uses sunlight to make its own food. If there is no light it can absorb food from the water. Some of its close relatives have no green pigment and, like animals, feed on other living things. The larger algae are green, brown, or red. Many seaweeds seen on the shore are brown algae. Most seaweeds are red algae. They can live in relatively deep water. Algae reproduce in various ways. Some just divide; others make tiny sex cells that join to form a new generation. Some reproduce by shedding spores into the water.

The variety of algae

Algae show an enormous range of sizes and shapes, almost as though they had experimented with every way of putting cells together to make a larger body. Shown here are just a few of the green algae. *Spirogyra* (photo below) is an alga that forms a green scum on fresh water. The microscope shows that its filaments are made of cells joined end to end. It can be recognized by its chloroplast, which forms a spiral. *Staurastrum* (1) lives floating in lakes. It is single-celled and the cell wall is spiky. In *Eudorina* (2), a freshwater form, 32 cells live together in a ball of jelly. Species of *Cladophora* (3) are found in fresh water and on the seashore. Attached at one end, the filaments of cells branch and have a wiry feel. Some grow 1 m (3 ft.) long. *Acetabularia* (4) lives on rocks on the lower shore and consists of a single cell as long as your finger. *Codium* (5) is a seaweed about 30 cm (12 in.) long. It has a mass of filaments interwoven into branching tubes with a feltlike feel. The sea lettuce, *Ulva* (6), has fronds with a double layer of cells.

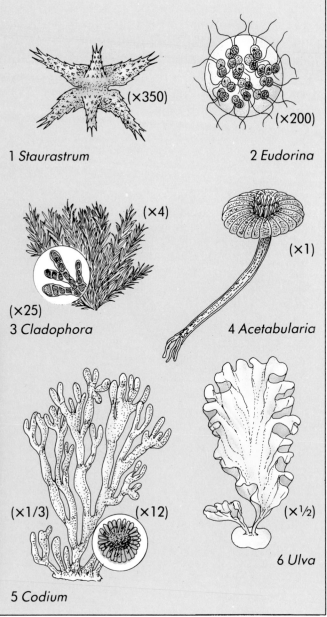

1 *Staurastrum* (×350)

2 *Eudorina* (×200)

3 *Cladophora* (×4) (×25)

4 *Acetabularia* (×1)

5 *Codium* (×1/3) (×12)

6 *Ulva* (×½)

Fungi

Fungi differ from true plants in having no green pigment and not making their own food. They need to get food from outside their bodies. Many are parasites of plants, and include serious pests of crops. Some live on the bodies of live animals. Large numbers get their food by breaking down the remains of dead animals or plants. Some form associations with plants, especially algae, from which both benefit.

There are about 100,000 species of fungi.

Some, like the yeasts that ferment and help us make beer, wine, and bread, are single-celled. But many fungi are much larger. Often the main part of a fungus remains unseen, and consists of a mass of threads, or hyphae, underground. We notice the fruiting bodies that grow from these threads because they grow above ground. They release spores in thousands which are carried through the air. Some find a suitable spot to grow. Some fruiting bodies are small, and rarely seen. Others are the familiar mushroom and toadstools.

◄▼ It is in the fruiting bodies of fungi that spores are produced. A variety is shown below. In the puffball (photo) the spores are puffed out through the hole on top when it is disturbed.

Mushroom life cycle

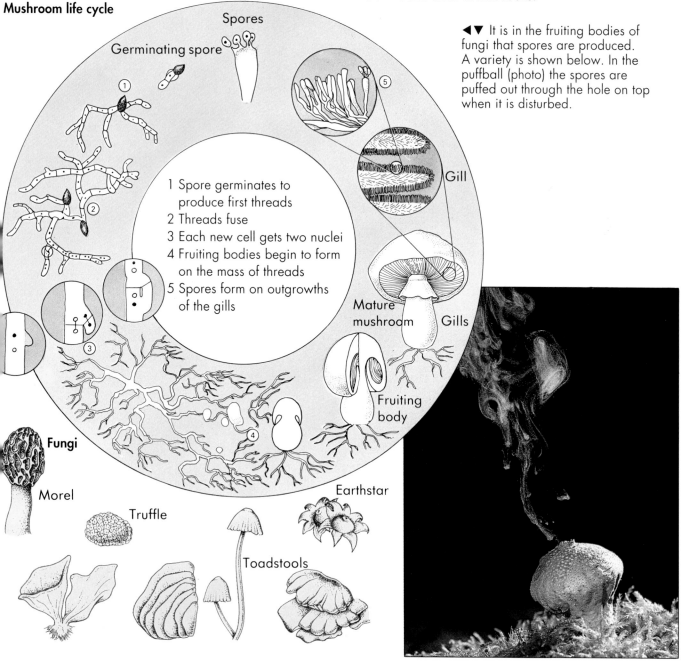

Spores

Germinating spore

1 Spore germinates to produce first threads
2 Threads fuse
3 Each new cell gets two nuclei
4 Fruiting bodies begin to form on the mass of threads
5 Spores form on outgrowths of the gills

Gill

Mature mushroom

Gills

Fruiting body

Fungi

Morel

Truffle

Earthstar

Toadstools

False morel   King Alfred's cakes   Bracket fungus

Liverworts, mosses, and ferns

Mosses and liverworts are low-growing plants that often form mats or cushions. They are found mainly in damp and shady places, although some mosses can stand being dry for months. They have no true stems or leaves, nor do they have proper roots. They absorb water and foods from below through hairlike cells. They have no special tubes for carrying water and food, and this limits their size.

Mosses and liverworts have two stages to their life cycles. One stage produces male and female cells. These unite to produce the next stage, which is a spore-bearing plant. The spores give rise to a new sexual generation. Liverworts get their name from their sexual stage. This is often a flat but fleshy plant body shaped like a liver. Other liverworts have flattened "leaves" growing from a central body. Mosses always have leaves carried on a stem. The sporophyte, or spore-bearing part, of these plants grows on the sexual part. It consists of a foot, a stalk, and a spore capsule.

Ferns also have sexual and spore-bearing stages. The sexual stage is small, short-lived, and rather like a flat liverwort. The familiar large ferns such as bracken are spore-bearing stages. Under the fronds you can find brown patches, which are where the spores form. Ferns have tubes to take water and food through their bodies, and can grow much larger than mosses. They have roots and leaves which are often complex fronds. These are able to catch enough light even in shady places. Like mosses and liverworts, ferns are commonest in warm countries. Ferns are most abundant in Southeast Asia. A few grow tree-sized.

Horsetails have an upright stem with rings of side branches. Few species of these ancient plants survive. The spore-bearing parts form "cones" at the tip of the shoots. Like ferns, they eject their spores clear of the plant.

Although most are not conspicuous there are about 9,500 different kinds of moss, 6,000 liverworts, and 12,000 ferns.

Moss life cycle

▶ Spore capsules grow from a moss plant (1) and (2). When mature, their caps come off, exposing a ring of "teeth" (photo below). This opens in dry air and lets out spores (3). A spore grows into a green thread (4). Moss plants bud from this (5). Male parts (antheridia) and female parts (archegonia) develop on the plant (6). Male cells swim to fertilize the egg (7), which divides (8) to form a new sporophyte.

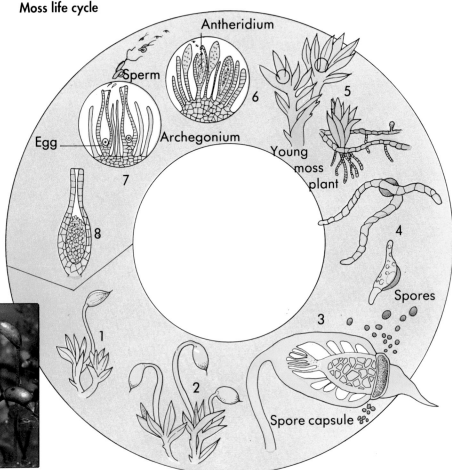

▲ A damp southern beech forest in New Zealand. Here mosses, ferns, and lichens flourish, many of them growing on the trees. Ferns and mosses need a film of water for their male cells to swim through in order to reach the female cells. Damp conditions, as here, are ideal for reproduction. Ferns and mosses thrive in such wet places, but cannot succeed in drier areas, such as deserts.

◄ The British soldier lichen is two plants combined: an alga and a fungus. Often the same sort of alga also lives free, but the fungus needs the alga to live. One way lichens reproduce is by throwing off balls of fungus threads enclosing algal cells. These bundles give a dusty look to the soldier lichen. An association such as this, where two species combine, is called symbiosis.

▼ A lichen is made of an alga cell surrounded by fungus threads.

Alga _____ Fungus

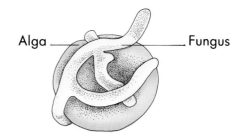

Plants and pollen

Pollen is a type of plant spore that contains male cells. It provides a good means for dispersal, even in dry conditions. The first plants to produce pollen were the gymnosperms, or "naked-seed," plants. Another type of spore, which develops into female cells, grows on the plant inside its "spore case." Once male cells from pollen fertilize the female cell, it develops into an embryo still inside this case. The embryo with its case is called a seed. In gymnosperms, seeds have no other cover. In flowering plants, which are called angiosperms, seeds are wrapped in another layer. The whole thing is known as a fruit. Plants have many ways of dispersing their seeds.

SPOT FACTS

• A cone on the cycad *Encephalartos caffer*, from Africa, can weigh up to 42 kg (over 90 lb.).

• The largest seed is that of the coco-de-mer from the Seychelles. It can weigh 18 kg (40 lb.).

• The *rafflesia* is the plant with the biggest flowers. The flowers measure nearly 1 m (about 3 ft.) across, and have a foul smell.

• Nasty-smelling flowers are called carrion flowers, because they attract insects such as flies, which feed off, and lay their eggs in, rotting flesh.

• Grasses cover about one quarter of the Earth's land surface. All grasses are flowering plants.

• The only cycad native to North America is found in the sandy woods of Florida. It is a source of *sago*, from which bread and other starchy foods are made.

• Many insects are discriminating pollen carriers. They fly from flower to flower of the same species.

• Conifers use the wind for pollination. They produce pollen in such large quantities that pine forests can be enveloped in a haze of pollen in spring.

• Because plants make food by photosynthesis, they are the foundation of the food web. All animals rely directly or indirectly on plants for energy.

• In order to collect nectar from tubular flowers, hummingbirds have extremely long tongues. They are also able to fly backward.

• Conifers include the largest organisms on the planet, the record being held by a specimen of the Coast redwood called General Sherman. It is 84 m (275 ft.) high.

• Seeds of the Oriental lotus have been known to germinate 3,000 years after dispersal.

• A great many early medicines – among them aspirin, digitalis, opium, and quinine – were derived from plants.

• Fragrances of flowers are designed to attract pollinators; they are produced by minute quantities of volatile oils in petals.

• What insects see when they look at flowers is quite different to what we see because their eyes are sensitive to ultraviolet light. This means that some petals with patches not visible to the human eye may look spectacular to a butterfly.

• Many flowers vary in color according to the pH (acidity) level of the soil. Hydrangea flowers may be pink on clay soils and blue on lime soils.

• To prevent self-pollination, many plants produce only male flowers on one plant, and only female flowers on another.

• Insects are not the only flower pollinators. Many flowers are pollinated by birds, such as the hummingbird, and some are pollinated by bats.

• Plants go to great lengths to distribute their seeds. Many are carried in the digestive system of birds. Other plants rely on wind and water to carry seeds.

Primitive gymnosperms

The first of the gymnosperm group of plants lived over 220 million years ago, before the time of the dinosaurs. They had developed two big advantages over plants that had gone before. They had pollen that could be carried long distances, and they had seeds. In a seed an embryo plant can remain dormant until conditions are right for it to grow.

The first seed plants we know of were rather fernlike. Later they grew as large trees. Most of the living gymnosperms are conifers such as fir trees. But there are still a few examples of old-fashioned types of gymnosperm in isolated parts of the world.

The cycads have spreading palmlike leaves forming a crown on an unbranched trunk. The trunk has a large soft core, or pith. Some cycads are quite short. Others grow up to 20 m (over 60 ft.) high. Long ago they were common, but now only about 100 species survive. Spores are formed in huge cones on the top of the trunk.

There is only one kind of ginkgo, although we know about others through fossils. It grows as a tall narrow tree up to 28 m (92 ft.) high. It is planted in cities because it resists pollution.

The gnetophytes are another small group of odd seed plants. They include the shrubby *Ephedra*, from which the drug ephedrine can be extracted. Oddest of all is *Welwitschia*, a rare plant from Namibia in southern Africa. It is a desert plant, good at preventing the loss of precious water. It is woody and has just two straplike leaves. These grow throughout the life of the plant, and may curl and split to produce a large twisted plant.

▼ This cycad grows in Mexico. It has edible seeds. Cycads look rather like palms, but belong to a separate group of plants more closely related to conifers. They were most successful 150 million years ago, but a small number of species still live scattered through the tropics. The stems of some can be used to make sago.

The living fossil

Leaves and catkins of a ginkgo tree. The ginkgo seems almost unchanged from those that grew 200 million years ago. Now it grows wild only in a remote part of China. Pollen is produced on the catkins and carried by the wind. The female parts sit at the ends of special shoots.

The conifers

There are about 550 species of conifer. Conifers are most common in the cooler parts of the world. Some live at the limits of tree-growing zones on mountains or toward the poles. Conifers are mostly upright plants and are woody shrubs or trees. They include the tallest plants on Earth, the giant redwoods.

The cones that give these trees their name come in two forms, male and female. These are usually on the same tree but on different shoots. The male cones are usually small, but the mature female cones can be woody structures up to 40 cm (16 in.) long in some species. A few species, such as junipers and the yew, produce a fleshy cone. The red yew "berry" is attractive to birds. This helps distribute seeds, as the birds carry the seeds away from the tree.

Conifers usually have needlelike leaves. Some, such as cypresses, have short scaly leaves around their shoots. The monkey puzzle tree from South America has huge spiky scales. Conifer needles have a hard outer skin and are a good shape for retaining water. Conifers can live in dry conditions, as in the cold northern winter, or on poor sandy soils. Most conifers keep their leaves all year.

Conifers can form dense forests. Because they let little light through to the ground, these forests may have little undergrowth. But many fungi live in the ground in association with their roots. Some appear as mushrooms.

Conifers grow fast in cool climates and on soils that are no good for other crops, so they are much used in forestry. Their wood, called

softwood, can be used as timber and also for papermaking.

The Norway spruce is a common Christmas tree. Europe's tallest native tree, at up to 54 m (177 ft.), it may be 30 years old before it starts making cones. The age of many conifers can be judged by the number of rings of branches. There is usually one for each year. But in species such as the Scotch pine the older trees often lose their lower branches.

▼ The young cones of a Norway spruce. The pine cone consists of special spore-bearing leaves, or scales, formed into a tight cluster. The spores are protected by the casing of the cone's hard scales until they are fully ripe. The female cones produce the seeds, the male cones the sperm cells, or pollen. The male cones are noticeably smaller than the female cones.

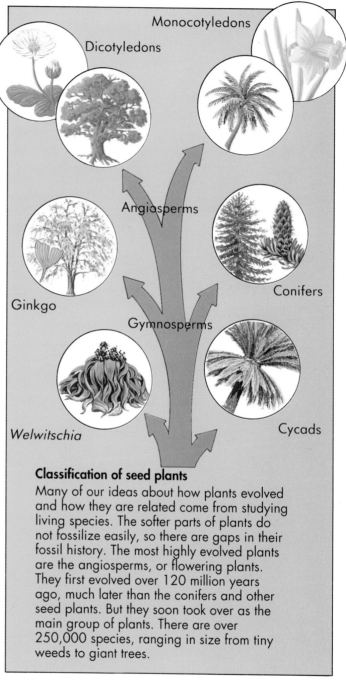

Classification of seed plants

Many of our ideas about how plants evolved and how they are related come from studying living species. The softer parts of plants do not fossilize easily, so there are gaps in their fossil history. The most highly evolved plants are the angiosperms, or flowering plants. They first evolved over 120 million years ago, much later than the conifers and other seed plants. But they soon took over as the main group of plants. There are over 250,000 species, ranging in size from tiny weeds to giant trees.

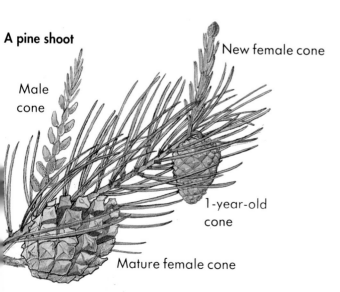

A pine shoot

Male cone

New female cone

1-year-old cone

Mature female cone

◄ (Far left) The Monterey pine is found in just one small coastal area of California. This isolated population is descended from pines left behind as conifers retreated north after the last Ice Age.

◄ In conifers, reproduction is a slow process. In the pine, winged pollen from male cones is carried by the wind to a young open female cone. It may take a year for the pollen to germinate and grow to fertilize the egg cell. The female cone closes up after pollination. Seeds mature slowly, and a female pine cone may be three years old before it finally releases the ripe seeds.

The flowering plants

Nine-tenths of that part of the world that is covered with vegetation is covered with flowering plants. Many are important to us for food, drugs, or timber or as ornamental plants.

There are two main types: dicotyledons, which start life as embryos with two leaves, and monocotyledons, which have a single leaf in the embryo. Some of the other differences between the groups are shown below. More than two-thirds of flowering plants are dicotyledons.

The dicotyledons include many small flowers such as primroses and violets, shrubs such as hazel, and trees such as elm and oak. The monocotyledons are mostly low-growing. They include the grasses, lilies, orchids, and palms.

Some flowering plants are pollinated by wind. Most have their pollen taken from flower to flower by insects. The flower itself, with its colors and its nectar, attracts the insects which do this important work for the plant.

Dicotyledons	Monocotyledons
 Embryo: Two seed leaves. Endosperm (extra food) present or absent.	Embryo: One seed leaf. Endosperm often present. 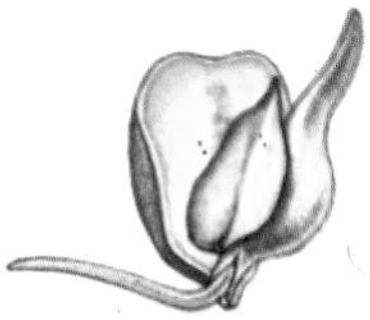
 Roots: The first root often persists and becomes a strong taproot, with smaller secondary roots.	Roots: The first root soon disappears, to be replaced by a branched fibrous root system. 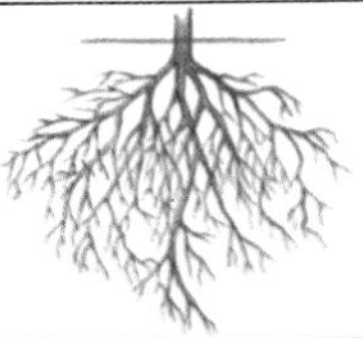
 Growth form: May be a herb, that is, a low-growing plant, or may be a woody shrub or tree.	Growth form: Most are herbs. A few, such as palms, are treelike. 
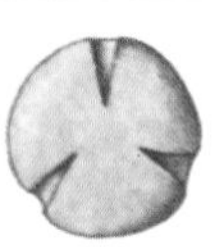 Pollen grain: Usually has three furrows or pores.	Pollen grain: Usually a single furrow or pore. 
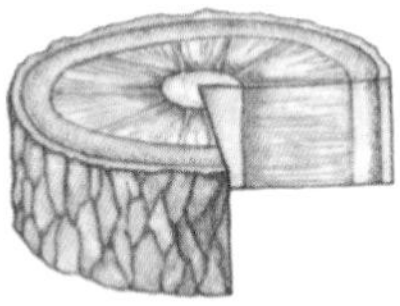 Vascular system: Tubes for conducting water and food are in a ring around the outside of the stem.	Vascular system: Tubes for conducting water and food are scattered across the width of the stem. 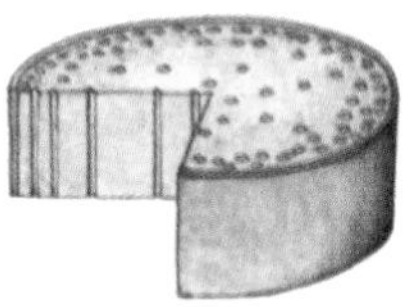
 Leaves: Usually broad, and with the veins forming a network. May be divided or compound.	Leaves: Usually long and narrow, often sheathing the stem. The veins run parallel to the long axis. 
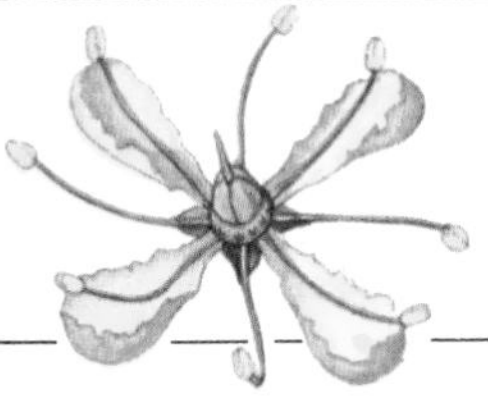 Flowers: Parts are typically arranged in fours or fives.	Flowers: Parts are usually arranged in threes or multiples of three. 

The seeds of flowering plants are enclosed within a fruit. Sometimes this is hard, like the acorn. Sometimes it is fleshy, like the plum or blackberry. Like flowers, fruits often attract the attention of animals, which carry seeds away from the parent plant to grow.

Some flowering plants are protected against attack by producing strong-tasting or poisonous substances in their leaves. These can stop animals from eating them.

Parts of a flower

Carpels

Stamens

Petals

Sepals

▲ The hogweed is a member of the umbellifer family, which includes the carrot and parsley. The flower head is up to 20 cm (8 in.) across and is made up of many small flowers, the outside ones larger. Many of this family have subdivided leaves.

▲ The magnolia is an ancient type of flower. Scientists believe that all flowers developed from this type. It has female parts, the carpels, and male parts, the stamens, as well as petals and sepals. In more advanced plants there may be fewer parts. Some may be lost, or fused to make complex new structures, as in orchids.

▲ *Kerria* is a flowering shrub from China often planted in gardens. It belongs to the rose family and the flower has five petals. The leaves are simple, that is, undivided.

Animals without backbones

The earliest known fossil animals, from 650 million years ago, are soft-bodied jellyfish. By 600 million years ago we know there were many other animals, including some with shells or hard skins, but they did not have backbones. These were invertebrates. Since then, animal bodies have become more complicated and efficient. Better senses and brains have evolved. But many of the 39 main groups, or phyla, that we know today already existed 600 million years ago. Some animals have hardly needed to change since then, perhaps because they fit so well with their surroundings. This group includes the highly successful insects.

SPOT FACTS

• The largest animal without a backbone is the giant squid. It can reach 20 m (over 60 ft.) in length and can weigh up to 2,000 kg (4,400 lb.).

• The Great Barrier Reef off the coast of Australia is made of coral. It is 1,900 km (1,200 mi.) long, the largest structure built by living things.

• Spiders and scorpions are generally beneficial as they eat insect pests, and only a few species are dangerously poisonous to humans.

• Spider silk has been successfully used for making the cross-hairs in optical instruments.

• Many roundworms such as the pinworm and the hookworm are parasites of plants and animals; some cause major health problems such as trichinosis and elephantiasis.

• Young slugs usually have well-developed shells, but these are either lost or are reduced to a small remnant by adulthood.

• Although most mollusks have poor vision, cephalopods such as octopuses have eyes with lenses, retinas, and other features amazingly similar to those of vertebrates.

• Some oysters care for their young inside their mantle cavity. They can switch sex back and forth from male to female.

• Some crustaceans are huge. The spider crab can have a leg span of 3.6 m (12 ft.).

• Gastropods range in size from being barely visible to weighing, in the case of the sea slug, three or four times the weight of an average domestic cat.

• Sea cucumbers are echinoderms that live on the sea floor. They are close relatives of the starfish. When threatened, they can eject a sticky net to entangle their aggressor.

• Some small parasitic insects are less than 0.025 cm (less than 0.01 in.) long when fully grown, whereas at least one fossilized species related to modern dragonflies is known to have had a wingspan of more than 60 cm (24 in.).

• Some species of cicada can take up to 17 years to reach maturity. Once an adult, they may live for only a month.

• Parasitic bumblebees invade the nests of similar species; they often kill the resident queen and force the workers to raise their young.

• Worker bees feed developing larvae as much as 1,300 times a day.

• A dragonfly's eye has 30,000 facets.

• The goliath bird-eating spider of Guyana has a body length of about 9 cm (about 3.6 in.).

• Small spiders can travel for hundreds of miles in the wind suspended from a "parachute" line.

• The bite of the black widow spider is rarely fatal in humans. It can, however, cause fainting and difficulty in breathing.

Classifying the animals

Animals are classified into phyla depending on their body plan. For example, all soft-bodied worms with bodies consisting of a series of rings, or round segments, are put in a phylum called the Annelida. The Mollusca have soft unsegmented bodies protected by one shell or two. Animals in most phyla have a head and a tail. However, the Cnidaria and Echinodermata are symmetrical around the vertical axis.

One stage up from the Protozoa, which have a single cell, sponges have more than one layer of cells but no specialized nerve or muscle cells. The higher animals have three cell layers. These animals have "organs" containing cells that are specialized for a particular job.

Almost all invertebrate phyla contain species that live in the sea. Fewer species are adapted to fresh water. Only the jointed-legged animals such as insects and spiders are common on land.

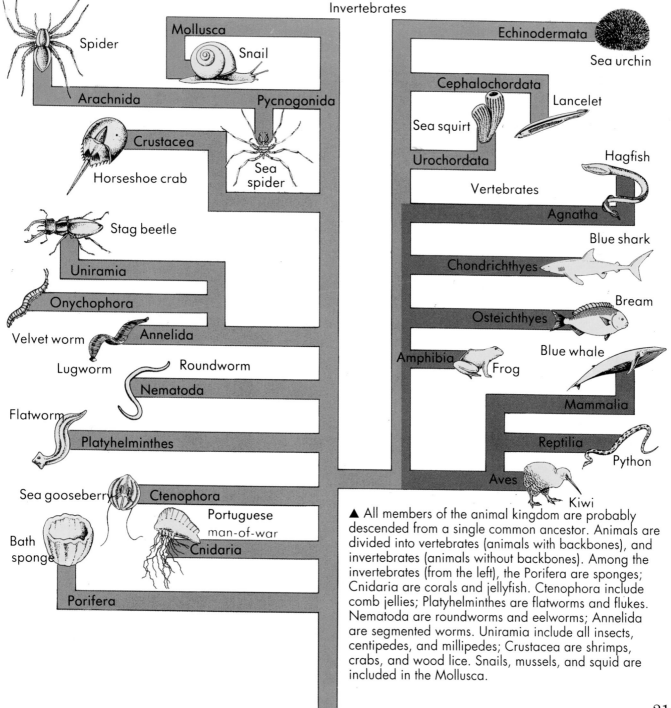

Invertebrates

Spider — Arachnida
Mollusca — Snail
Pycnogonida
Crustacea — Horseshoe crab
Sea spider
Stag beetle — Uniramia
Onychophora
Velvet worm — Annelida
Lugworm — Roundworm
Nematoda
Flatworm
Platyhelminthes
Sea gooseberry — Ctenophora
Portuguese man-of-war
Bath sponge — Cnidaria
Porifera

Echinodermata — Sea urchin
Cephalochordata — Lancelet
Sea squirt
Urochordata
Vertebrates
Hagfish
Agnatha
Blue shark
Chondrichthyes
Bream
Osteichthyes
Amphibia — Frog
Blue whale
Mammalia
Reptilia — Python
Aves — Kiwi

▲ All members of the animal kingdom are probably descended from a single common ancestor. Animals are divided into vertebrates (animals with backbones), and invertebrates (animals without backbones). Among the invertebrates (from the left), the Porifera are sponges; Cnidaria are corals and jellyfish. Ctenophora include comb jellies; Platyhelminthes are flatworms and flukes. Nematoda are roundworms and eelworms; Annelida are segmented worms. Uniramia include all insects, centipedes, and millipedes; Crustacea are shrimps, crabs, and wood lice. Snails, mussels, and squid are included in the Mollusca.

Simple animals

Sponges live on the seabed. They have a central cavity, and draw water into this through many small holes in their bodies. It is sent out of one large hole at the top. On the way, food is removed. Many sponges have hard crystals of silica in their bodies. The bath sponge has tough fibers instead.

Cnidarians include corals and jellyfish. They have circular bodies. There is a single opening into the body cavity where food is digested. They catch prey using the stinging tentacles around this mouth. Some corals form huge colonies which secrete a stony skeleton to live in. Coral reefs form in tropical seas and take centuries to build up.

Several kinds of animals are called worms. Annelids include earthworms, which do useful work in loosening the soil. Many annelids live in the sea. Some burrow, for example lugworms. Others live in tubes. Some walk on the bottom on fleshy "legs." Leeches have suckers and jaws to feed on the flesh or blood of other animals.

Flatworms are another important group. Some glide over the bottom of the sea or fresh water, but many kinds are parasites. These live in other animals, including humans. They include tapeworms and liver flukes. Roundworms have no segments, and are long and narrow with pointed ends. They have very tough skins. Many live in the soil, but some are parasites on plants, animals, or humans.

▲ Part of the Caribbean seabed. The leaflike sea fan and the brain coral in the foreground are cnidarians, as is the branching hydroid colony. A flat red encrusting sponge can be seen. Tube-shaped yellow sponges point upward. None of these animals can move around. They sieve out tiny food particles in the water.

▼ An earthworm is made up of a series of rather similar segments, but the hearts and sex organs are toward the front of the animal. It does not have complex eyes, but simply cells on its back that are sensitive to light. Other cells respond to chemicals or touch. Worms swallow soil as they burrow and digest decaying plant fragments in it. Earthworms burrow in soils all over the world, and need to keep their slimy skins damp.

▼ The fanworm is a segmented worm that lives on the seabed. It builds a tube and lives inside. The head is crowned with a ring of tentacles covered in tiny beating hairs, or cilia. The fan of tentacles is spread from the end of the tube and used as a gill to get oxygen from the water. It also traps tiny pieces of food. These are passed down by the cilia to the mouth.

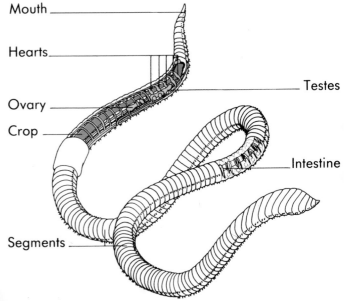

Mouth

Hearts

Ovary

Crop

Testes

Intestine

Segments

The mollusks

There are nearly 80,000 species of mollusks. They include snail-like forms with a single coiled shell; clams and mussels with a pair of shells; and the octopuses and squids, in which the shell is internal or absent. Land snails are usually plant eaters. Some of the sea snails are meat eaters. Bivalves such as mussels are filter feeders and are able to stay in one place and let their food come to them. Squid and octopuses are fast-moving, active hunters. They have efficient bodies, good eyesight, and large brains. They are the most intelligent of all the invertebrates. Most puff out a cloud of "ink" into the water when they feel threatened.

▲ The head of a land snail. The tentacles bear sense organs including eyes. The mouth contains a long tongue covered with sharp teeth. It is used like a file to rasp at the plants the snail feeds on. Snails are the only land mollusks. They glide on a carpet of slime. The shell gives protection from enemies and from drying out.

▲▼ A lesser octopus catches a crab. The arms grab prey, which is killed by a bite from the beak. Octopuses spend most of their time on the sea bottom. Squid and cuttlefish (below) are good swimmers. They take water into the mantle cavity for the gills. They can swim by jet propulsion by shooting water out of their "siphon."

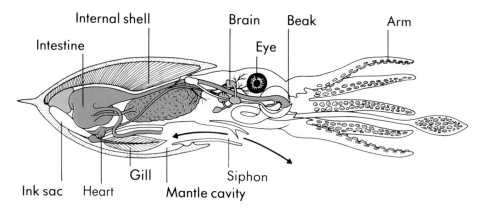

Internal shell   Brain   Beak   Arm

Intestine   Eye

Ink sac   Heart   Gill   Siphon   Mantle cavity

Insects

There are more species of insect than all the other kinds of animals put together. Over one million have already been named. Scientists believe there may be even more waiting to be described. Insects live in almost every place except the sea. Every food seems to have some kind of insect that eats it. What is the secret of insects' success?

An insect's body is divided into three main parts: the head, thorax, and abdomen. The head has eyes and antennae. The thorax has three pairs of legs and usually two pairs of wings. Being able to fly is a big advantage. Insects have a hard outer skin that also acts as a skeleton to support them. This skin is quite waterproof, and helps insects succeed on land.

The disadvantage of such a skin is that it cannot grow. To grow, an insect must shed its skin, and expand while the new one is soft. There are two main ways that insects grow up. Some insects, such as grasshoppers and cockroaches, hatch from eggs as small versions of adults. But they lack wings, and are unable to breed. They molt several times before becoming adults. Other insects, such as butterflies, flies, and wasps, hatch from the egg as a wriggling larva quite different from the adult. This feeds, molts, and grows. Then it goes into an immobile stage, the pupa, in which its body is totally reorganized. From it, the insect emerges as a fully formed winged adult.

Some insects such as bees, ants, and termites are social animals with highly organized nests. Some insects damage plants and stored crops. Others carry disease. But many are important to agriculture in pollinating crop plants.

The variety of insects. The fighting male rhinoceros beetles (1) are 16 cm (6 in.) long. A hunting wasp (2) has caught a fly. An assassin bug (3) feeds on a caterpillar pierced by its sharp jaws. A southern hawker dragonfly (4) lays her eggs on a waterlogged branch. Black garden ant workers (5) tend aphids. They protect them from enemies and in return "milk" them for their sweet honeydew secretion. The grasshopper *Oedipoda miniata* (6) is well camouflaged, but if it is disturbed it jumps (7), showing its bright wings, before settling and "disappearing" again. The monarch butterfly (8) makes long-distance migrations from America, where it feeds on the milkweed plant, south to Mexico, where it hibernates. In the spring it migrates north again.

1 Rhinoceros beetle

2 Hunting wasp

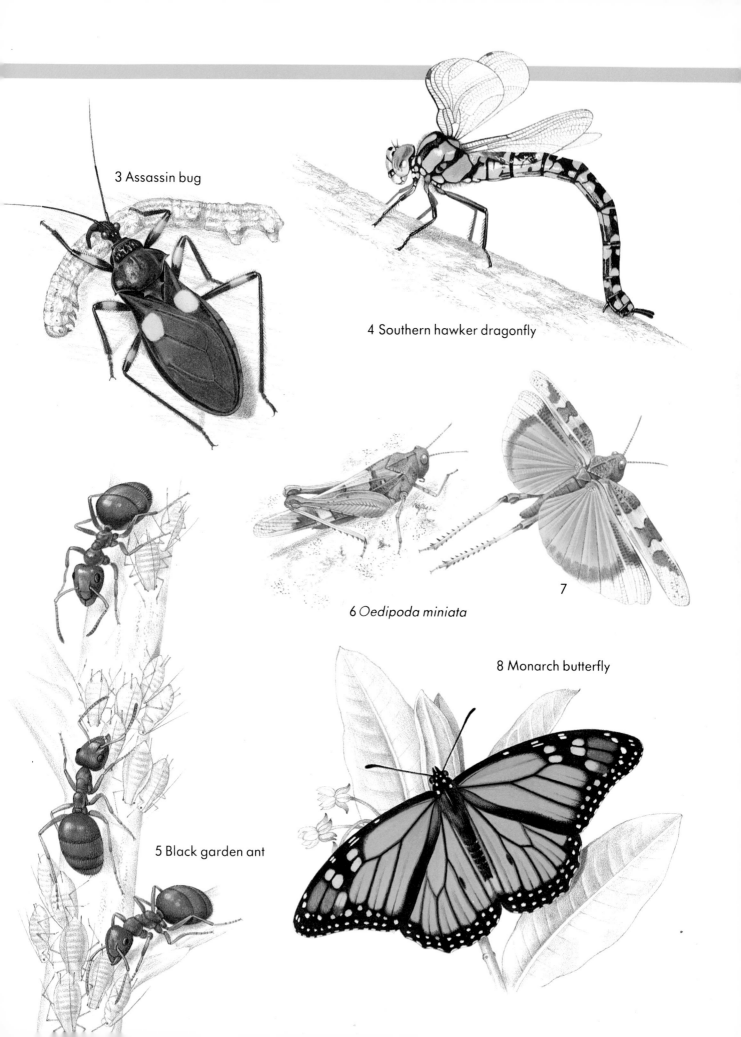

3 Assassin bug

4 Southern hawker dragonfly

6 *Oedipoda miniata*

7

8 Monarch butterfly

5 Black garden ant

Other jointed-legged animals

Apart from insects there are several other groups of jointed-legged animals, or arthropods. Centipedes get their name from their supposed 100 legs. In fact most have only about 40, although one species has 354. Centipedes are hunters. They have poison claws around the mouth. With these they catch insects and small worms. The largest tropical species, 33 cm (13 in.) long, can catch mice. Millipedes are a separate group. Instead of having one pair of legs on each segment like centipedes, they have two pairs. They are harmless. Most feed on decaying plants. They live in damp, dark places or in the soil. Some tropical species are up to 30 cm (12 in.) long. Some have 750 legs, but none has 1,000, as their name would suggest.

Arachnids include spiders, mites, and scorpions. Scorpions live in the warmer parts of the world. Like other arachnids they have four pairs of walking legs. In front of these they have large pincers which they use to seize and crush prey. They have a long tail with a sting at the end that curves over the back. This is used for defense and to subdue prey. Few species have a sting that is dangerous to humans. Scorpions mate after a "dance" in which the pair hold pincers. Scorpion young are born alive and climb onto the mother's back. There they stay until after their first molt. Like other arthropods they molt their outer skeleton as they grow and are vulnerable until the new, larger coat has grown hard.

▼ Soldier crabs march across an Australian shore. Crabs are found on shores worldwide, and also in the depths of the sea. A few species in the tropics even live on land. Most crabs are hunters or scavengers that feed on dead animals on the seabed. Some tropical species pick up pellets of mud and sieve them for food.

Spiders are all equipped with poison fangs to paralyze or kill prey. Not many species are dangerous to people. Most specialize in hunting or trapping insects. Spiders have special silk-producing glands. The silk they produce can be stronger than steel of the same thickness. The silk is used for safety ropes or parachuting. In many spiders it is used for making webs to trap prey, or wrapping prey when caught. But some spiders do not use webs. Jumping spiders and wolf spiders pursue their prey on the ground. Crab spiders sit on flowers that match their color, and ambush prey.

▼ (Below right) A male orb-web spider courts a female by twanging her web. Female spiders are usually bigger than males, and males approach with care to avoid being mistaken for food. (Bottom) *Argiope bruennichi* spins a strong, sticky orb web to catch prey. Like many orb-web spiders it spins a new web each night.

Most arthropods are land animals, although one large group, the crustaceans, has largely remained in the water. This group includes the crabs and lobsters and the many kinds of shrimp. There are almost 40,000 species. Crustaceans are an important food, both in fresh water and seawater, for many other animals. Some, such as the krill of the cold seas, may live in shoals of millions.

A lobster has a segmented body. The head has eyes and two pairs of antennae. The thorax and a long abdomen or tail follow behind. Each segment has a pair of legs. Those near the mouth work as jaws. Those under the thorax are walking legs. Those under the abdomen are swimming legs. A crab is much the same, but its small tail is tucked under the body. Other crustaceans show variations on this body plan. Some are quite extreme, such as the barnacle.

Orb-web spiders

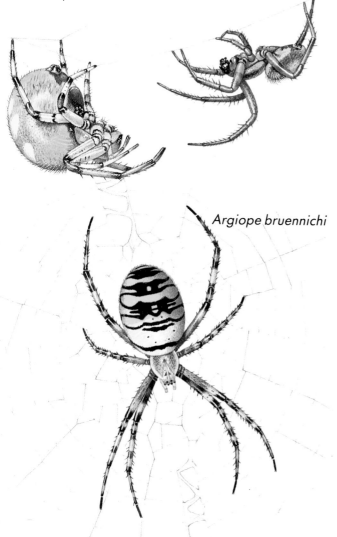

Argiope bruennichi

Echinodermata

The echinoderms are invertebrate animals that live in the sea. The phylum includes starfish, sea urchins, sea cucumbers, and brittle stars. These spiny-skinned animals are built on a symmetrical five-rayed pattern. They have complicated bodies, with a skeleton of hard plates in their body walls. They have a unique system of water canals that works the "tube feet" on which these animals move. The photograph above shows a starfish using its tube feet to grip prey as it tries to pull open a scallop shell to eat its flesh.

Fish, amphibians, reptiles

Nearly all big animals are vertebrates, or backboned animals. They are a very successful group. Their body chemistry works fast. Many are very active. They have developed body shapes and limbs to move in water, on land, and in the air. They have the biggest brains and include the most intelligent animals. Fish, amphibians, and reptiles are "cold-blooded." Birds and mammals maintain their own body temperature.

Vertebrates first evolved in the water. Later they colonized the land. Over hundreds of millions of years they have developed efficient bodies, brains, and methods of reproduction, putting them in a position of dominance over all other life on the planet.

SPOT FACTS

• The smallest fish is the Philippine dwarf goby. Adults may be no more than 0.8 cm (0.3 in.) long.

• A large python may go without food for over a year without harm.

• The largest snake accurately measured was a reticulated python 10 m (over 30 ft.) long, from Southeast Asia.

• The gila monster is the only poisonous lizard in the United States. Its bite can be fatal to humans.

• The paradoxical frog tadpole is 25 cm (10 in.) long. The adult is under 7 cm (3 in.) long.

• A cod lays 6 million eggs in a single batch. Only a handful survive to become adults.

• The largest rays are known as devilfish, sea bats, or mantas; they attain a weight of more than 1,360 kg (over 3,000 lb.).

• The lampfish is so called because the Native Americans of Alaska use the dried fish, which is very oily, as a lamp. They push a piece of bark through the fish as a wick.

• The whale shark is the largest-known fish. Native to tropical seas around the world, the whale shark may attain a length of more than 15 m (50 ft.) and can weigh more than 18 metric tons.

• Male seahorses take charge of the eggs once they have been laid. They are placed in an abdominal pouch in which they remain until they hatch.

• Swordfish, if wounded and approached by a boat, have been known to drive their long beaks into solid-wood planking 5 cm (2 in.) thick.

• Tortoises are often long-lived; members of some species live for more than 100 years.

• Some snake venoms can have medicinal uses. They have been used as painkillers in cases of arthritis and cancer.

• The largest living lizard is the135-kg (300-lb.) komodo dragon.

• The most deadly of all fish venoms is found in the stonefish. It has been known to kill divers who have accidentally stepped on a stonefish.

• Monitors are among the oldest living lizards. They are related to the mosasaur, which grew up to 9 m (30 ft.) long and lived from 136 to 65 million years ago.

• The largest of all frogs is the African giant frog. It can grow as long as 66 cm (26 in.) and can weigh as much as 4.5 kg (10 lb.).

• The electric eel has electric "radar" to locate its prey. It can emit from 450 to 600 volts to stun its prey.

Vertebrate ancestors

Fishes were the first of the vertebrates. We do not know all the stages by which they developed from simple invertebrates. Their immediate ancestors probably looked much like the living lancelet. Rather fish-shaped, the lancelet has blocks of muscle running down each side of its body. There is a tail fin, but no others. A mouth near the front lets in water which is passed out through gills. These obtain oxygen, and also filter food out of the water. This animal does not have a true backbone but has a stiffening rod, a notochord. Vertebrates have this during development before the backbone. The lancelet has no brain, skull or jaws, or true heart. The most primitive vertebrates, such as lampreys, still have no jaws. They have larvae that look very like lancelets. From beginnings such as this, true fish developed, with a proper backbone. They also developed fins, and a head with senses, a brain, a skull, and jaws.

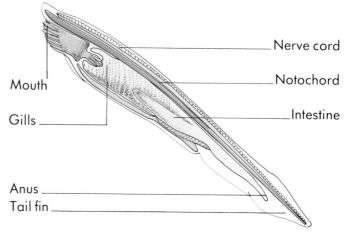

▲ The lancelet shows many of the features that might be expected in an ancestor of the vertebrates. Similar animals lived 550 million years ago.

▼ The red sea squirts have tadpole-shaped larvae with notochords, suggesting they are related to vertebrates.

Fish

There are two main groups of fish. The cartilaginous fish include the sharks and rays. They have rough skins with scales like little sharp teeth. In many ways they seem to be "old-fashioned" fish, and some have hardly changed for millions of years. But there are still over 600 species, among them the biggest fish of all, the plankton-eating whale shark.

The bony fish are much more numerous. There are 20,000 kinds. They range from the large to the tiny. They are often streamlined to move fast through water, but there is an enormous variety of shapes and ways of life in this group. Eels, sea horses, catfish, flatfish, pike, puffer fish, and flying fish are just a few of the shapes bony fish can take.

One small group of bony fish are called lobe-finned fish. They have paired fins with bones. Included in this group are the coelacanth and the lungfishes. They are survivors of a group that was successful a long time ago. It is thought all land animals are descended from their early relatives.

Most bony fish belong to the group called ray-finned fish. Their paired fins are supported by thin bony rays. They breathe through gills. A single slit behind the gill cover lets out water that has come from the mouth via the gills.

Fish eat a huge variety of foods. Some strain the water for plankton. Others eat plants or snails. Some, such as piranhas, are fierce hunters with sharp teeth. Bony fish may have teeth not just on their jaws, but on the tongue, roof of the mouth, or throat lining too. Most bony fish lay large numbers of eggs which are fertilized externally and hatch as tiny larvae. Only a few kinds take care of their young.

▶ A great white, or man-eater, shark homes in on a bait. The teeth work like a saw blade. Sharks have stiff pelvic fins that act like wings to give them "lift" in the water, as they have no swim bladder. They are not good at sudden turns or braking. The tail is not symmetrical. Its shape helps to lift the rear end as well as driving the fish forward. The mouth is below the head, and the large snout holds a good sense of smell. There is a row of separate gill slits on each side of the neck. The backbone is made of cartilage, or gristle, and not bone.

▼ A male sunfish stands guard with fins erect. The spines in the front part of the dorsal fin can be raised or lowered. The pelvic fins are close under the pectorals, giving good turning and braking ability. The swim bladder helps give the fish buoyancy. Bony fish such as this maneuver with ease.

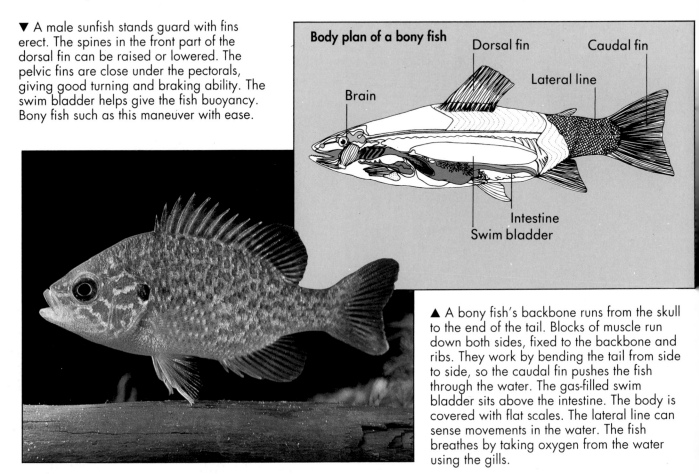

Body plan of a bony fish

Brain • Dorsal fin • Lateral line • Caudal fin • Intestine • Swim bladder

▲ A bony fish's backbone runs from the skull to the end of the tail. Blocks of muscle run down both sides, fixed to the backbone and ribs. They work by bending the tail from side to side, so the caudal fin pushes the fish through the water. The gas-filled swim bladder sits above the intestine. The body is covered with flat scales. The lateral line can sense movements in the water. The fish breathes by taking oxygen from the water using the gills.

Amphibians

Amphibians include frogs, toads, newts, and salamanders. They are mostly land animals, but their lives are tied to the water. They lay their eggs in water and these develop into tadpoles, which swim and have gills to breathe. Later the tadpoles develop legs, lose their gills, and develop lungs. At this stage, as tiny frogs or newts, they move to the land. Even then, they mostly inhabit damp places, because their smooth, slimy skins are not waterproof. The advantage of these skins is that they can be used for breathing, which means they are able to help out the simple lungs.

Over 350 million years ago some lobe-finned fish lived where pools sometimes dried out. They had to wriggle to the next pool to survive. Gradually some became better adapted for making these trips, and the fins became more like legs. Air breathing became important. Amphibians had arrived. The head of amphibians differs from that of fish. Amphibians have eyes with eyelids and tear glands, and blink to clean the eye. Fish have internal ears and hear well. On land something more is needed to pick up sounds in the air. Amphibians have eardrums at the back of the head. For frogs, sound is important for attracting a mate.

Newts probably look most like the early amphibians. Frogs and toads are relatively new, but they are very successful, especially in the tropics. There are 3,500 species, compared with about 500 for other amphibians.

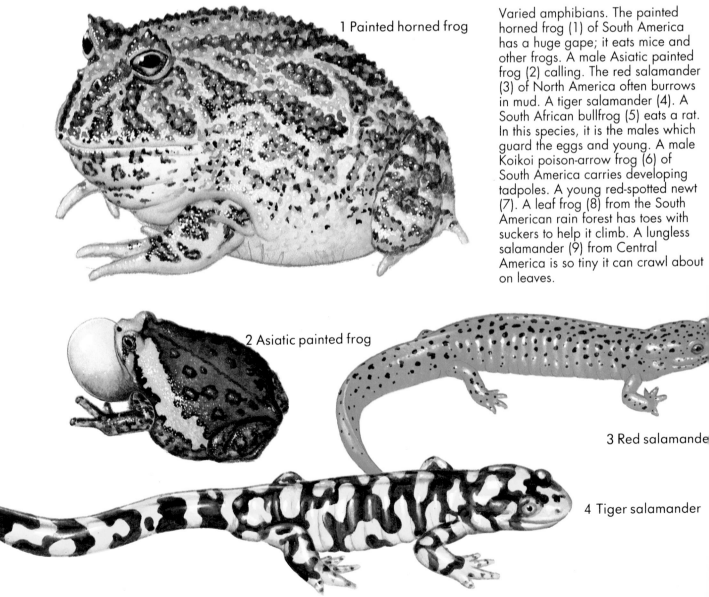

1 Painted horned frog

Varied amphibians. The painted horned frog (1) of South America has a huge gape; it eats mice and other frogs. A male Asiatic painted frog (2) calling. The red salamander (3) of North America often burrows in mud. A tiger salamander (4). A South African bullfrog (5) eats a rat. In this species, it is the males which guard the eggs and young. A male Koikoi poison-arrow frog (6) of South America carries developing tadpoles. A young red-spotted newt (7). A leaf frog (8) from the South American rain forest has toes with suckers to help it climb. A lungless salamander (9) from Central America is so tiny it can crawl about on leaves.

2 Asiatic painted frog

3 Red salamande

4 Tiger salamander

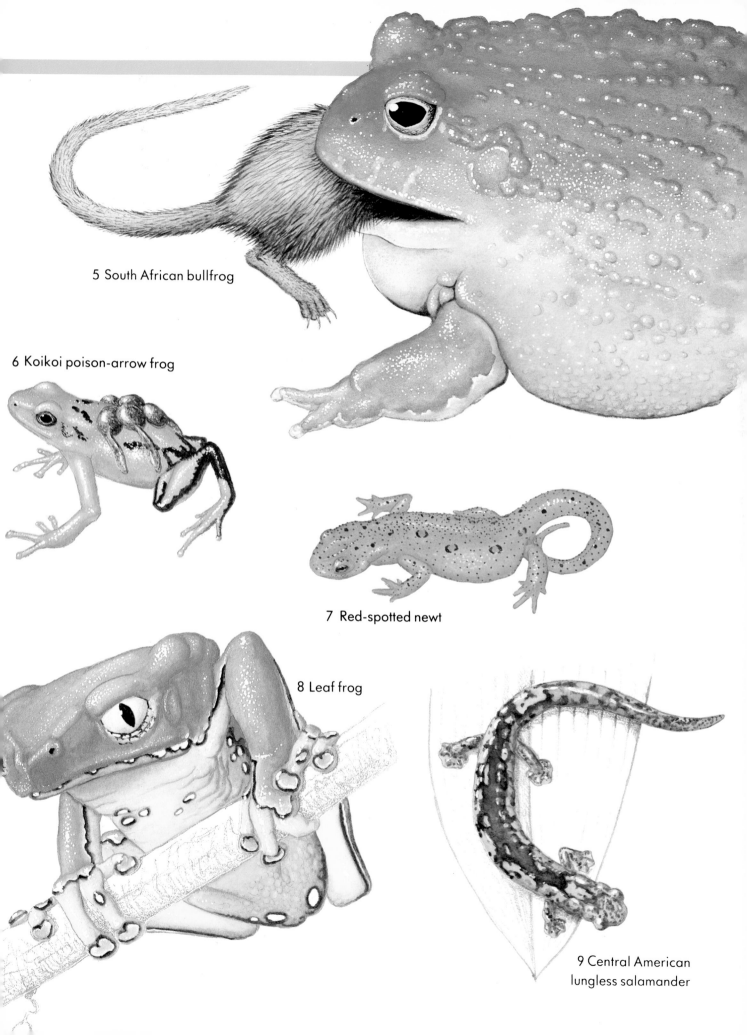

5 South African bullfrog

6 Koikoi poison-arrow frog

7 Red-spotted newt

8 Leaf frog

9 Central American
lungless salamander

Reptiles

Turtles, crocodiles, snakes, and lizards are all reptiles. Reptiles evolved from amphibians and in several ways are better at the job of living on land. Among other things their legs are generally longer and stronger, and their lungs are more efficient. But perhaps the two most important improvements reptiles have made are in their skins and in their breeding. Reptile skins are dry, scaly, and quite waterproof. Reptiles do not need damp places. Many live in deserts. They do not need water to breed either. Unlike the soft amphibian egg that has to develop in water, a reptile egg has a hard or leathery shell, and is laid on land. From this hatches out the young, which is a small replica of its parents.

Most reptiles do not look after eggs and young. Eggs are laid and left to hatch on their own. A reptile cannot incubate its eggs, as it is "cold-blooded." This means it depends on its surroundings for warmth, rather than generating heat in its body. Reptiles are most common in the tropics. Many bask in the sun to raise their temperature for activity.

Tuftles are one of the most ancient reptile types, but also perhaps the oddest. The shell roughly corresponds to the ribs and scales of other reptiles. Land tortoises are usually slow-moving plant eaters. Sea turtles and fresh-water turtles are good swimmers and are mostly hunters. They must surface to breathe. Crocodiles also swim well, and are the largest living reptiles and fierce hunters.

Snakes and lizards are the most numerous reptiles. There are about 3,000 kinds of each, compared with 250 species of turtle and 25 of crocodile. Most lizards are small and feed on insects. They are found in every habitat from the tops of trees to burrows underground. Most snakes are hunters and swallow prey whole.

1 American alligator

2 Gavial

3 Black caiman

4 Flap-necked chameleon

5 Gopher tortoises

6 Green turtle

7 Cuban ground boa

A selection of reptiles. An American alligator (1) sits on her nest mound. Several kinds of crocodile are known to guard nests and young. The gavial (2) is a narrow-snouted fish-eating crocodile from India. It grows to over 6 m (20 ft.) long. The black caiman (3) lives in South America. A flap-necked chameleon (4) catches a fly. Chameleons can shoot out their tongues in a fraction of a second. These have sticky grasping tips. Gopher tortoises (5) are good burrowers. These two are mating. A green turtle (6) feeds on sea grasses. The Cuban ground boa (7) is one of the smaller constricting snakes. It swallows prey whole, and digests it slowly.

Birds and mammals

Birds and mammals are the most highly evolved vertebrates. They are "warm-blooded" and able to keep themselves at a constant temperature in a variety of surroundings. So they can always be active, even in cold conditions that would stop a reptile's body from working. A disadvantage of this is that a warm-blooded animal needs more energy, and must feed regularly. Insulation is also needed on the outside to conserve heat. Birds and mammals have the largest brains among the vertebrates. Some are very intelligent. They are also very good at looking after their young.

SPOT FACTS

• A hummingbird may beat its wings up to 80 times a second.

• The longest time a mammal spends growing in its mother before birth is the 21 months of the Indian elephant.

• In flight, the heart of a small bat beats 1,000 times a minute.

• At up to 33 m (100 ft.) long, the blue whale is the biggest living animal.

• The kiwi of New Zealand lays the largest egg for its size of any bird. It may be one-fifth the weight of the mother.

• For lightness, the bones of most birds are hollow rather than filled with marrow, and are connected to a system of air sacs dispersed through the body.

• Some birds, such as pelicans, kingfishers, woodpeckers, and jays, are completely devoid of feathers when hatched.

• Of the nearly 10,000 species of birds known since historical records have been kept, at least 75 have become extinct.

• Birds inhabit every continent and almost every island in the world and are adapted to virtually every environment.

• The peregrine falcon and the common barn owl nest on every continent except Antarctica.

• In monotremes, the egg-laying mammals, mammary glands evolve equally in both sexes.

• The smallest bird is the bee hummingbird of Cuba. It is only 6.3 cm (2.5 in.) from bill tip to tail tip.

• The digestive system of carnivores is much less complicated than that of herbivores. This is because their digestive system does not have to break down the cellulose in plant matter.

• The male duckbilled platypus is one of very few mammals capable of producing venom, from a small gland on the rear ankle.

• Chimpanzees are born with 65 percent of their adult brain capacity, humans with only 25 percent.

• Only mammals have hair. It can take various forms such as fur, wool, bristles, and quills.

• In the manatee the hindlimbs have been lost. Whales have vestigial limbs beneath the skin.

• Cats and dogs can only sweat through the soles of their paws.

• All dogs except the African hunting dog have five toes on the forefeet and four toes on the hindfeet.

• The rabbit population of New Zealand all bred from seven rabbits, which were introduced into the country near Invercargill, apparently about 1860.

• The only true wild horse still in existence is Przewalski's horse, now native only to parts of western Mongolia.

Birds

There are some 8,500 species of birds, and they live everywhere from the tropics to the Arctic and Antarctic. Birds have a body covering of feathers to keep them warm, but still have scales on their back legs like their reptile ancestors. Their front legs have evolved into wings, but still show the same basic arrangement of bones as a reptile.

Birds lay hard-shelled eggs, then sit on them to keep them warm until they hatch. Some lay eggs in little more than a scrape on the ground. Others build elaborate nests for the eggs and young. After hatching, some young are able to leave the nest immediately. They stay with their parents for a while for protection and to learn to find food. Other birds hatch blind and naked, and are kept warm and fed by the parents in the nest before fledging.

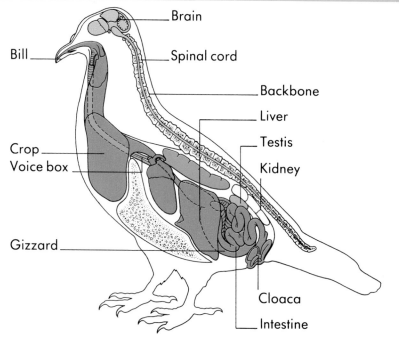

▼ A flock of galah cockatoos flies through the Australian countryside. The birds are a pest on crops, and their flying ability means they can travel considerable distances to find food. Their power of flight is also useful in helping them escape from their enemies.

▲ Anatomy of a bird. The heart is large, as might be expected in a very active animal. The brain is well developed. The backbone runs from the back of the skull to the base of the tail. A bird has no teeth. The gizzard helps to grind up the food that is swallowed.

Bird contrasts

Considering the advantages of flight it is surprising how many groups of birds have become flightless. Ostriches, cassowaries, and rheas are all good runners, but cannot fly. Penguins' wings have become adapted for "flying" through the water rather than air. The biggest flying birds weigh about 15 kg (33 lb.). They include the mute swan, the great bustard, and the Andean condor. The wandering albatross, which spends most of its life airborne gliding across the oceans, has, at 3.5 m (11 ft.), the longest wingspan.

At the other extreme is the tiny bee hummingbird, which weighs only 1.6 g (0.056 oz.).

Hummingbirds can hover in one spot to feed on nectar. Other birds including sunbirds and some small parrots like this form of food. A huge number of birds are insect eaters. They may have slender bills to pick insects off leaves. Some have other techniques, for example swifts and swallows which pursue insects through the air using the beak as a net. Seed eaters include finches with triangular bills to match the size of seed they eat, pigeons that swallow grain whole, and parrots with powerful pincer beaks. The biggest parrot, the hyacinthine macaw, can easily crack Brazil nuts. Other birds kill large prey, or eat carrion or fish.

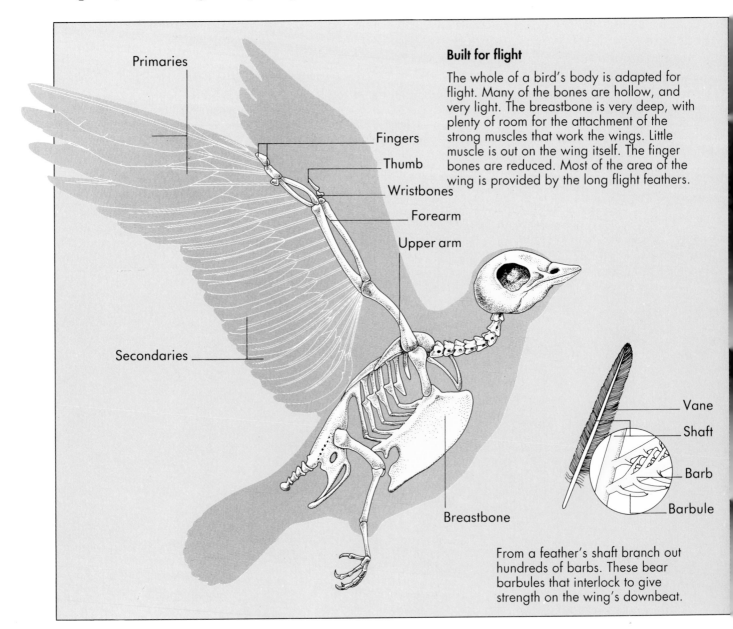

Built for flight

The whole of a bird's body is adapted for flight. Many of the bones are hollow, and very light. The breastbone is very deep, with plenty of room for the attachment of the strong muscles that work the wings. Little muscle is out on the wing itself. The finger bones are reduced. Most of the area of the wing is provided by the long flight feathers.

Primaries

Fingers

Thumb

Wristbones

Forearm

Upper arm

Secondaries

Vane

Shaft

Barb

Barbule

Breastbone

From a feather's shaft branch out hundreds of barbs. These bear barbules that interlock to give strength on the wing's downbeat.

Contrasting birds. The tufted duck (1) is a good diver, paddling with its feet in pursuit of small fish and insects. The whiskered tern (2) has a buoyant flight. It hovers above the water, then darts down when it spots a small fish. The marsh harrier (3) is a bird of prey. It has a sharp hooked beak, and sharp claws for catching prey. The ostrich (4) is the largest bird and is completely flightless, although it has quite large wings, which it uses in display. The legs are long and end in two hooflike toes. It has the biggest eyes among the vertebrates. The New Zealand South Island brown kiwi (5) is also flightless and has tiny wings. It is active at night and probes the ground for worms using its sensitive beak.

2 Whiskered tern

1 Tufted duck

3 Marsh harrier

4 Ostrich

5 South Island brown kiwi

Classifying the mammals

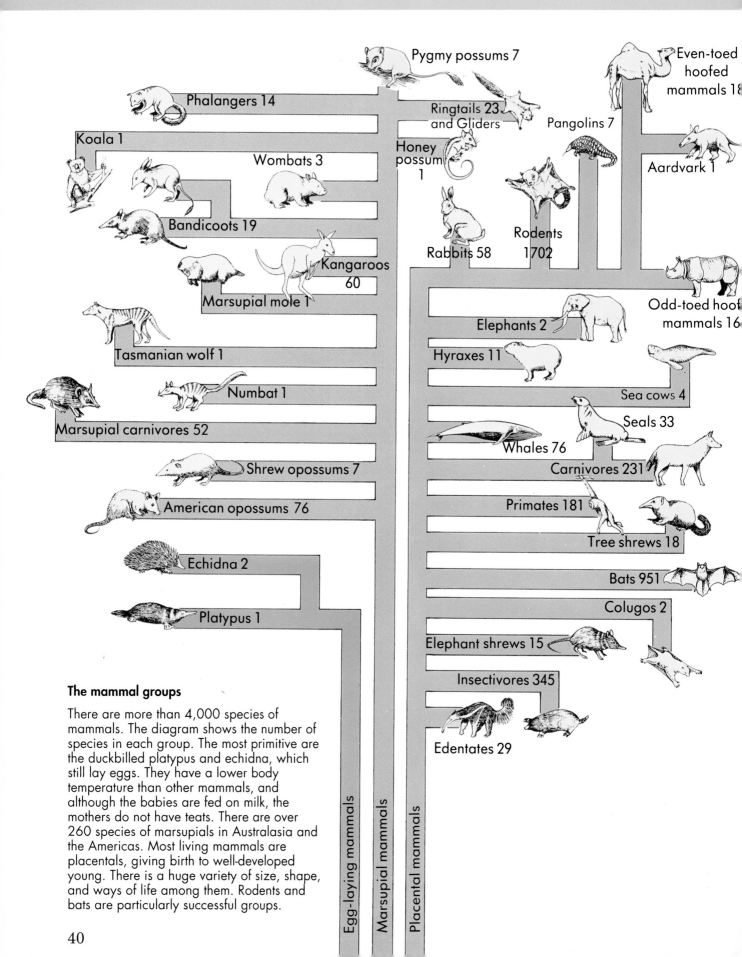

Pygmy possums 7

Phalangers 14

Koala 1

Wombats 3

Bandicoots 19

Ringtails 23 and Gliders

Honey possum 1

Even-toed hoofed mammals 18

Pangolins 7

Aardvark 1

Rabbits 58

Rodents 1702

Kangaroos 60

Marsupial mole 1

Tasmanian wolf 1

Numbat 1

Marsupial carnivores 52

Shrew opossums 7

American opossums 76

Echidna 2

Platypus 1

Elephants 2

Hyraxes 11

Odd-toed hoof mammals 16

Sea cows 4

Seals 33

Whales 76

Carnivores 231

Primates 181

Tree shrews 18

Bats 951

Colugos 2

Elephant shrews 15

Insectivores 345

Edentates 29

The mammal groups

There are more than 4,000 species of mammals. The diagram shows the number of species in each group. The most primitive are the duckbilled platypus and echidna, which still lay eggs. They have a lower body temperature than other mammals, and although the babies are fed on milk, the mothers do not have teats. There are over 260 species of marsupials in Australasia and the Americas. Most living mammals are placentals, giving birth to well-developed young. There is a huge variety of size, shape, and ways of life among them. Rodents and bats are particularly successful groups.

Egg-laying mammals

Marsupial mammals

Placental mammals

Mammals keep a constant body temperature. Most have a coat of fur to help stay warm. With a thick coat a polar bear or arctic fox can live in the far north with temperatures at minus 40°C (-40°F). Mammals can live equally well in the tropics. Although the largest may not need fur for warmth, even elephants have some hair on the body. Fur can be raised or lowered to help adjust body temperature. Shivering, sweating, and panting can also play their part.

Placental mammals have babies born after a relatively long gestation period. Some, such as antelope, may be able to walk soon after birth. Others, such as mice or dogs, are quite helpless when they are born. In both cases the babies are fed on milk by the mother. Even after they are weaned many mammal young stay with their mother. They are able to learn while still getting protection. In some species it may be years before they are totally independent.

Placental mammals probably originated from ancestors that looked like shrews. From these beginnings the shapes of mammals' bodies have changed to produce good runners. There are also mammals which are good climbers swimmers, burrowers, and fliers. Body size ranges from that of the 150-ton blue whale to that of a bat which weighs only 1.5 g (0.05 oz.).

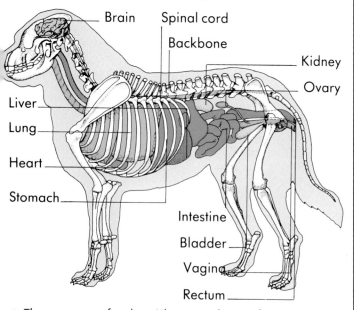

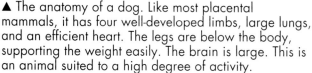

▲ The anatomy of a dog. Like most placental mammals, it has four well-developed limbs, large lungs, and an efficient heart. The legs are below the body, supporting the weight easily. The brain is large. This is an animal suited to a high degree of activity.

Marsupials

An eastern gray kangaroo of Australia and its pouched young. Like other marsupials, it is born after a very short gestation period and completes its development in the mother's pouch. The baby is smaller than your thumb when it is born. It is blind and has large front legs and small back legs. It crawls to the mother's pouch and there fastens on a teat. It stays in the pouch for 5 to 11 months until it is kangaroo-shaped and able to move outside. Marsupials have developed many different life-styles. Kangaroos themselves are like antelope in the way they deal with plant food. Their hopping gait is unusual, but can be fast. A large kangaroo may cover 9 m (30 ft.) in one bound. There are marsupial moles, anteaters, hunters, burrowers, and gliders that parallel mammals in other parts of the world. All have babies born as little more than embryos, which hold onto a teat. But not all marsupials have a real pouch.

Carnivores

Mammals of several different groups, or orders, are carnivores in the sense of being flesh eaters. But one particular order is known as the Carnivora, because nearly all its members specialize in a meat diet. These carnivores include the dogs, cats, bears, raccoons, weasels, and mongooses.

These animals have long, pointed canine teeth for biting to kill. The cheek teeth may be few in number, but have pointed ridges so that the top and bottom jaw can act together as scissors for slicing meat. Hunters need good eyes, ears, and nose for finding prey. Then they need speed to catch it. Many of the carnivores, such as dogs and cats, stand on their toes rather than the soles of the feet. Agility and being able to squeeze through burrows may also be useful, as is seen in some weasels and mongooses.

Meat can be quick to digest and nutritious. It can be roughly chopped and swallowed in lumps. Only a short digestive system is needed. Once it has had a good meal, a carnivore may not become hungry again for some while. In between hunts, it can afford to be lazy. A lion may have a good meal once or twice a week. It may sleep for 20 hours a day. It needs to do little until it is hungry again.

Other hunters include insect eaters such as shrews and bats, with small pointed teeth. Those that eat ants, such as pangolins, may do without teeth, but have long sticky tongues to pull in their meal. Fish eaters such as dolphins and some seals have pointed teeth to catch the slippery prey. They are streamlined for pursuing prey through the water and their limbs have been modified into flippers.

1 Smoky bat

3 Red fox

2 Killer whale

Animals that prey on others. The smoky bat (1) of South America feeds on small insects caught while flying. The killer whale (2) grows 8 m (26 ft.) long and can swim at 50 km/h (30 mph). It is one of the fiercest hunters in the sea. It catches seals and other whales. The red fox (3) kills small animals such as rabbits and voles, and may also scavenge, like the four shown here. It also eats worms. Like some other mammals, this species can vary in color. The black bear (4) may attack prey as large as deer, but is comparatively slow-moving. It also feeds on berries and other plants. The fat-tailed dunnart (5) is a mouse-sized marsupial that is a fierce hunter of insects. The Algerian hedgehog (6) eats insects and any other small animals it can catch. The false killer whale (7) catches squid and large fish with its sharp teeth. The South American sea lion (8) dives in pursuit of squid and large shrimp.

4 Black bear

5 Fat-tailed dunnart

6 Algerian hedgehog

7 False killer whale

8 South American sea lion

Herbivores

Several groups of mammals are adapted as herbivores, or plant eaters. The rodents, such as mice, squirrels, and beavers, have large teeth for gnawing at the front of their jaws, and a battery of chewing, grinding teeth along the side. As well as eating grass or leaves, many are able to gnaw into tough foods like bark or nuts. Most rodents are small animals. Rabbits and hares are not rodents, but they have the same types of teeth and feed in similar ways.

Most of the big plant eaters are hoofed mammals. There are two groups. The odd-toed hoofed mammals include horses, tapirs, and rhinos. Among the even-toed group are deer, camels, antelope, and cows. Some of these animals browse leaves and twigs from bushes. Giraffes reach high into trees with their tongue and lips. Other hoofed animals have broad lips to feed from the ground on tough grasses.

Many plants have little food value, so plant eaters have to eat a lot. They spend most of the day feeding. They must always be alert for enemies. The food needs a good chewing by the grinding teeth. The food is hard to digest, and so the digestive tract is long. Even-toed hoofed animals have complex stomachs where food ferments. They bring balls of food, the cud, back to the mouth for a second chewing. Odd-toed hoofed mammals do not chew the cud, but bacteria in the intestines help digestion.

1 Topi antelope

2 Plains zebra

3 European hare

44

Plant-eating mammals. The topi antelope (1) ranges widely through the grasslands of Africa. It likes fresh green grass. The plains zebra (2) also grazes on the African grasslands. The European hare (3) eats grass, but may nibble twigs and bark too. The Papuan forest wallaby (4) feeds on leaves. Wallabies and kangaroos have a large stomach in which bacteria help break down food. Rats, such as the African marsh rat (5), can often gnaw tough seeds. The rufous rat kangaroo (6) feeds on roots, tubers, and leaves.

5 African marsh rat

4 Papuan forest wallaby

6 Rufous rat kangaroo

Primates

The order of mammals we belong to is known as the Primates. It includes monkeys and apes. Most primates live in trees. They have eyes that face forward to judge distances as they jump and climb. They have good grasping fingers and toes, and have pads on the end of their fingers and nails rather than claws. Some South American monkeys, such as the brown capuchin, can grasp with their tail too. Many monkeys feed on both animal and plant food. The adaptations monkeys have for climbing also make them good at investigating their surroundings. They can look carefully, and use their hands to touch objects, or move or pull the objects apart. These adaptations were also important to our ancestors, allowing them to develop their intelligence.

Brown capuchin

Sooty mangabey

Night monkey

Part Two

Animal behavior

To survive, animals must have bodies built in the right way to fit the conditions in which they live. But this is not enough. They must do the right things too. They must have the right "behavior" to fit in with their surroundings and with other animals of their own kind. A deer may have long legs to escape from trouble, but if it does not run when danger threatens it will not survive.

In the fight for survival there is a constant contest between the hunters and the hunted. Some animals are in competition with others of their own kind for mates or food. In some species, individuals cooperate to bring benefits for all. All kinds of animals need to be able to breed and raise young. Whether they are trying to win a contest or mate, or to cooperate, animals often need to communicate with one another. Many have a "language" of signs and sounds that help them to get their meaning across.

◄ Worker honeybees tending the honeycomb. Bees are very successful social insects with well-developed ways of communicating with each other. In the comb, eggs are laid, larvae reared, and honey stored.

The fight for survival

All animals are engaged in a struggle for survival. Unless a species can breed successfully it will die out. Enough animals must survive to become breeding adults to keep the species going. They must find enough food for themselves and raise young. Meat-eating animals are engaged in a continual battle of wits with their prey. An animal needs to do the right things – to have the right behavior – if it is going to stand a chance of hunting successfully. The same is true of the animals that are hunted. They, too, have developed their own particular kinds of behavior to avoid becoming prey. But it is in the nature of things that these tactics do not always work. Some animals are caught and eaten, but hunters may miss a kill. Hunters are never so efficient that they wipe out their prey. Instead, there is a natural balance.

SPOT FACTS

• A leopard may be successful in only one in twenty of its attempts to catch prey.

• A pangolin may eat as many as 200,000 ants in one night, undeterred by the formic acid that ants use for defense.

• As flatfish move around the seabed, they may change color and pattern within seconds.

• Many bad-tasting insects get their taste from the plants they eat.

• When attacked, the bombardier beetle can shoot out a hot, pungent gas from its abdomen.

• In the nineteenth century, lions ranged through Europe and Asia; they were widespread in the grasslands of Iran and India. Today, they live wild only in Africa.

• The golden eagle may achieve a wingspan of up to 2 m (about 6.5 ft.).

• The largest crocodile is *Crocodylus robustus*, found in Madagascar. It can grow to 9 m (30 ft.), making it the largest living reptile.

• Ladybugs or ladybird beetles are so called because people in the Middle Ages regarded these pest-destroying creatures as instruments of benevolent intervention by the Virgin Mary.

• All birds of prey are carnivorous, except for one African vulture that feeds on palm nuts.

• Some brightly colored butterflies have toxic chemicals in their tissues, concentrated from poisonous plants they fed on as larvae.

• Several species of otherwise defenseless moths and flies avoid predation by birds by mimicking the buzzing sound of stinging bees.

• Some edible butterflies defend themselves by looking like poisonous ones. This is known as Batesian mimicry after Henry Bates (1825–1892), the English naturalist and explorer who discovered this phenomenon.

• Mimicry occurs among a vast array of different species, including insects and orchids, songbirds and hawks, and lizards and poisonous beetles.

• When fully extended for catching insects, the sticky tongue of the chameleon may be almost as long as its body.

• The snowshoe hare is pure white in winter, except for its black ear tips. In summer it is reddish-brown.

• Voracious sea hunters, the marlin and sailfish can attain speeds well in excess of 80 km/h (50 mph).

• The fur of the two-toed sloth can become encrusted in green algae. This helps it to camouflage itself in the trees.

Predator and prey

Predators are equipped in body and behavior for hunting. But hunting is not an easy process. Many meat eaters will also scavenge from the carcases of animals killed by others, or take prey away from a weaker animal. Jackals and hyenas are well known as feeders on carrion (dead meat) and as scavengers, but they also hunt for themselves. On the other hand lions, although capable of killing large animals, will sometimes catch small animals like rats, steal, or scavenge.

When predators hunt, they tend to take the easiest option. They go for prey that are young or sick or otherwise weakened. Oddly colored individuals may be singled out. Hunters chase lone animals, or try to separate one from a herd rather than be confused by a multitude. Dolphins or pelicans may use the opposite technique when they round up a shoal of fish into a small area. Here they rely on being able to scoop up many small prey when the prey are packed together.

When predators feed on prey larger than themselves, or in exposed surroundings, it increases the chance of success to hunt in groups. Lions, wild dogs, and hyenas on the African plains all tackle prey such as zebra or large antelope in this way. A zebra hunt may involve a group of a dozen spotted hyenas.

▼ A cheetah chases a Thomson's gazelle. Both are well adapted for running. A cheetah may reach 90 km/h (nearly 60 mph) over a short distance. The gazelle is not as fast but can run for a longer time. The cheetah attempts to catch its prey in a surprise dash lasting only about 20 seconds. It must get fairly close to its target before launching an attempt. If it fails it gives up and tries a different quarry. In these chases cheetah and gazelle are evenly matched.

▲ An orb-web spider bundles up a grasshopper that has blundered into its web. Such a spider uses little energy in hunting, as it simply waits for prey to come. But it has little control over the kind of meal that arrives. Spiders frequently use their poisonous bite to subdue large or dangerous prey as soon as it is stuck in the web. It is then rapidly wrapped up in silk and can be consumed at leisure. The main cost of this type of prey capture is the time and energy used in making the web.

Hunter and hunted

Predators

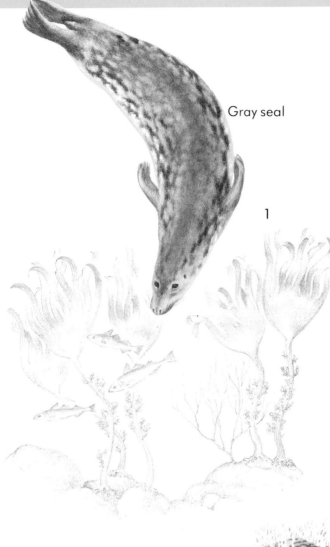

Gray seal

1

There are many techniques for catching prey. The first problem is to locate it. Birds, reptiles, and some fish and mammals rely on eyesight. Active hunters such as cats or owls have eyes that can look forward to judge distances. Other animals rely on the nose or ears. Insect-eating bats emit high-pitched sounds and listen for echoes reflected back from prey. Rattlesnakes have pits on the face which can detect the warmth of a mouse.

Some animals use speed to catch prey. Dolphins swim faster than the fish they capture. Cats may overtake their prey in a sprint. Dogs rely on sustained running over a long

Predators and prey. A gray seal pursues a fish through the water (1). An aardvark digs into a termite mound (2). After this it sits and licks up termites with its sticky tongue. A tiger catches and kills a spotted deer (3). Springing out in ambush it surprises the deer and seizes it by the neck, knocking it down. A throat bite then suffocates it. A herring gull acts as a "pirate," forcing a black-headed gull to drop the food it is carrying (4). A red-backed shrike has caught a lizard and impales it on its thorn "larder" (5). A noctule bat homes in on a moth that it has located by sound (6).

Spotted deer

3

Tiger

Aardvark

2

distance. Sparrow hawks fly along a hedge, hoping to surprise a small bird. Martens may chase squirrels at high speed through the trees.

Other hunters play a waiting game. A leopard waits on a branch for a victim to walk below. A kestrel hovers in the sky until it spots a vole, then swoops. Many spiders lie in wait. Some animals lure victims to a trap. Alligator snapping turtles have a wormlike mobile lure in their throats. Any fish trying to catch it is snapped up by the turtle. Many stealthy hunters rely on camouflage to keep them from being noticed. The praying mantis looks like a leaf until it shoots out its legs to grab a victim.

4

Herring gull

Black-headed gull

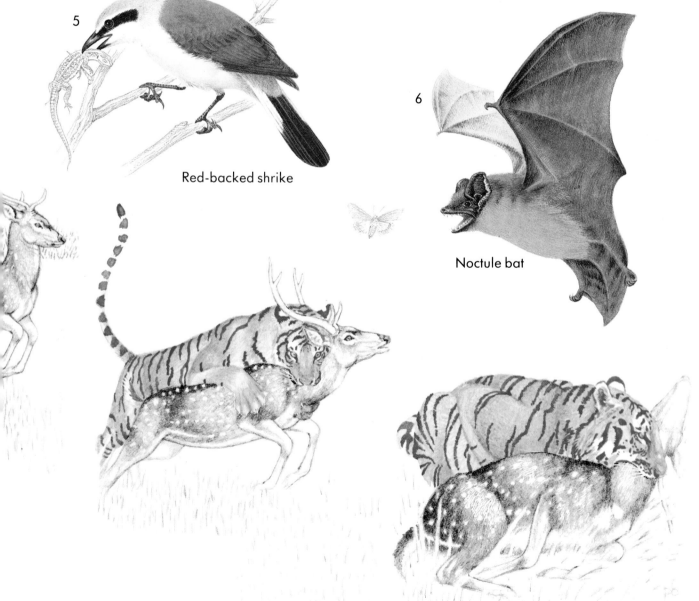

5

Red-backed shrike

6

Noctule bat

Defensive ploys

For some animals the first reaction to danger may be to dive for cover. A rabbit runs for its burrow. A worm goes underground. A crab scuttles under a rock. But some animals are too large to burrow. Deer and horses rely instead on swift running to evade capture. They may appear to ignore enemies, but if a predator comes within a certain critical distance, they instantly take flight.

Some animals carry their own protective covering with them. A snail can retreat into its shell when threatened. Clams can clam up. Many turtles can pull their head and legs into their shell. Some tortoises have hinged shells, and can shut them like trapdoors, so only hard shell is visible.

Some mammals carry strong protection. Armadillos have horny plates on their backs, and pangolins have overlapping scales like a pine cone. They curl up when frightened to protect their soft underside. These scales are really specialized hairs. Hedgehogs have another form of protective hair, their spines. If they are frightened, they erect their spines and curl into a ball. Porcupines have longer spines. Some crouch with erect spines if molested, but several kinds react to danger more actively. They stamp, grunt, and rattle their quills. If this does not deter an attacker, they turn and run backward to stab it, and perhaps leave quills in it. These are barbed and painful.

Many animals achieve safety in numbers. In a large herd or shoal there are more eyes watching for danger, and less chance of being taken by surprise. A herd moving about can also confuse a predator and upset an attack.

▶ A frightened three-banded armadillo curls up into an armored ball with no exposed soft parts. Some other species of armadillos are unable to make a good ball, and retreat into a burrow, plugging it with their backs.

▼ A European grass snake turns upside down with an open mouth, shamming death. Many predators are geared to attack only living prey. An apparent corpse will not be attacked. American opossums are famous for using this ploy.

▼ A herd of springbok flee from a predator. Some jump high in the air as they run. The erratic movements may confuse a predator.

Mimicry

Two butterflies of similar appearance but belonging to different species. The *Papilio dardanus* female (below) is a tasty species, but *Danaus chrysippus* (above) is poisonous and foul-tasting so it is avoided by predators. *Papilio dardanus* gains some protection by resembling the bad-tasting species. Predators learn to avoid butterflies that look like this.

For this sort of mimicry to work, the harmless mimic must not be too common compared to its dangerous model. If it is common, predators may not have enough opportunity to learn to avoid this pattern.

There are other groups of animals in which mimicry is found. Hover flies have no sting but are colored in such a way that they mimic wasps and bees that have stings. One large species of hover fly is actually found in several forms that mimic different species of bumblebee. King snakes, with no venomous bite, mimic the colors of the very poisonous coral snakes that live in the same region.

Danaus chrysippus

Papilio dardanus

Defense warnings and bluffing

Some animals deal with danger by distracting an enemy's attention. The peacock butterfly has colored "eyespots" at the edges of the wings. These grab a bird's attention, and when it tries to catch the butterfly it snaps at these, leaving the important parts undamaged and the butterfly able to fly away. Some fish have eyespots near their tails, and shoot away in the opposite direction to that expected.

Another way of confusing an enemy is to have bright colors that are only visible when the animal is moving. Some frogs have bright colors on the inside of their legs. If an enemy approaches too closely they jump, showing a flash of color which disappears when they land. Some butterflies have brightly colored underwings that serve the same purpose.

Some animals defend themselves with almost nothing but bluffing. Some harmless snakes puff themselves up, hiss, and lash their tails. Lizards such as chameleons also inflate and hiss. This may intimidate an attacker. Mammals that are afraid of attack often erect their fur so they appear much bigger than normal. They stand up tall to add to the effect. The fish known as puffers swallow water or air when attacked and quickly change into a prickly ball which it is almost impossible to grasp.

Some animals are equipped with weapons in the form of poison or a foul taste. Many newts,

frogs, and toads have poison glands in their skin which make them repulsive to attackers. But it is useful to be able to signal that they are dangerous before they are attacked and perhaps hurt. Many of the most poisonous species are brightly colored. Black and red or black and yellow are common warning colors. Caterpillars may have stinging hairs, like tiger moths. Or they may taste exceedingly nasty, like the cinnabar moth caterpillar. Some snakes such as cobras and coral snakes can defend themselves with a venomous bite. In all these cases colors are used as warnings. Cobras add to the effect with a hood on their throat. Rattlesnakes use the rattle on their tail.

▶ The spectacular lionfish of the tropical seas is easily spotted. The spines on some of its fins contain venom glands, and can deliver a powerful sting. Hence its other name, firefish. After an encounter with a lionfish, predators soon learn to avoid it.

◀ The frilled lizard of Australia has no special weapons. But when an enemy confronts it, it erects the big ruff of skin around its neck. It suddenly appears huge, and can scare off the attacker by pure bluff.

▼ (Below right) The African foam locust is red and black as a warning. If it is attacked it secretes a foul-tasting foam from between the joints of its body.

▼ Poison-arrow frogs produce a deadly poison in their skins. As the name suggests, this can be used for tipping arrows. But its real purpose is for defense. Poison-arrow frog species come in many bright colors.

Camouflage

Camouflage is any coloring that helps an animal fit in with its background or disguise its shape. The simplest camouflage is perhaps the dull browns or grays seen in many mammals. These do not attract the eye but blend in with many backgrounds. Some animals, though, are colored and patterned in detail to match the place where they are usually found. The match may even extend to shape. A leaf insect, or a stick insect on a twig, is very difficult to pick out from its background.

For camouflage to work, the right behavior is essential. The more an animal can freeze into stillness, the safer it is likely to be. Some caterpillars mimic little twigs. With the feet at the end of their body they grasp a twig. They hold themselves out at just the right angle and look like smaller twigs.

Even if an animal matches its background its shadow could be a giveaway, showing that a solid body is there. It is useful to eliminate or reduce the shadow. Flatfish and some flattened lizards leave no room below for one. Many animals, such as antelope, are colored above, but white or much lighter below. This softens the shadow and helps disguise the body.

Another form of camouflage seen in animals is disruptive coloration. At first sight this may not seem much like camouflage, as the animal may be quite vividly colored in contrasting stripes or patches. It works by distracting the eye from the whole animal. One or two parts may be noticed, but the eye sees disconnected blobs rather than a whole shape.

Birds such as plovers have this form of camouflage, and can be very difficult to pick out on a nesting beach. When resting, many moths have contrasting patches that split up their shapes. The okapi has a plum-colored body and contrasting striped legs. In the dark forest it may be hard to see. The zebra's stripes cut across the outline of its body. Although it seems so vivid, at a distance on the African plain it may not be noticed.

◀ The frogmouth is related to the nightjars, and like them hunts at night. During the day it relies on its camouflage for protection. Its colors blend in with tree bark. If it is disturbed it keeps perfectly still with its beak pointing upward, which aids the disguise.

▶ Three examples of camouflage. The sargassum fish (1) lives among the floating seaweed in the Sargasso Sea. Not only do its colors blend with its surroundings, but so does its shape. It even has spots and bobbles that mimic the little bladders on the seaweed. (2) Wild boars and their young in shady woodland. Wild boar adults are a dull brown all-purpose camouflage color, but during their early life the piglets are striped. This pattern matches the dappling found on the woodland floor. (3) Katydids are a type of grasshopper. Many are well disguised to fit in with their surroundings. Here a Central American katydid climbs among the leaves. In both color and shape it is a very good match for the leaves on which it is sitting.

2

3

The mating game

To reproduce successfully, most animals must first find a mate. Turtles and sea lions normally spread out over huge areas of sea and converge on particular beaches for breeding. Other species advertise for a mate by calling, as do many birds or frogs, or have colors to attract the opposite sex. Once located, a mate may be apprehensive of a stranger. Courtship may be needed to create confidence so that the pair will accept one another. Many different types of courtship can be seen. Some involve lengthy rituals, and often the males put on a display to attract the females. In some animals, one male and one female stay together, and may share the care of the young. In many kinds of animals, males mate with many females, and the mother is left to look after the young.

SPOT FACTS

• The size of a male northern fur seal's harem depends on his size and strength. A big male may collect as many as 40 females.

• Gannets pair for life, which can be up to 15 years.

• In a pack of African hunting dogs, it is only the leading female of the pack that breeds.

• The long black tail of the male East African widow bird, or whydah, attracts females from distances of over 1 km (0.6 mi.).

• Many bats mate before hibernation, but the eggs are not fertilized until the spring.

• Bowerbirds take their name from the elaborate structures they make as part of their courtship ritual. The gardener bowerbird builds a structure like a wigwam as much as 1.6 m (over 5 ft.) across. It has a low entrance, in front of which is a "garden" of bright objects and flowers.

• The great horned owl is a dedicated parent. In the northeastern United States where the winters are very long, they nest in the late spring. Often the incubating bird is covered in snow.

• The lyrebird is so-called because the tail of the male, which is raised forward over the back during courtship displays, resembles a lyre, with broad, curved outer feathers and threadlike inner feathers.

• Before leaving its nest, the Egyptian plover completely covers the eggs or chicks with sand to protect them against the intense heat of the desert.

• The ruffed grouse is famous for its springtime "drumming," a sound produced by the males beating their wings against the air as they stand erect. The sound carries a great distance.

• The female praying mantis sometimes eats the male's head before mating has finished.

• A female domestic dog can easily attract males from nearly a mile away by powerful chemical communicators known as pheromones.

• Male cockroaches flap their wings and reverse into objects behind them when they sense the female's pheromone. Eventually the males reverse into a female cockroach, and mating takes place.

• The female crocodile will often carry her newly hatched young in her immensely powerful jaws.

• Male spiders may attract a female's attention by plucking her web, but they have to take care to avoid being eaten.

• In kingfishers and mongooses, the young may continue to live with their parents to help with the next brood.

• The calf elephant suckles its mother's teats for five years before weaning.

Courtship

Many fish have good color vision and use colors in their courtship. The male guppy has brighter colors and bigger fins than the female, and spreads the fins in front of her. In many fish the male is equipped for display. In lizards, too, the male tends to be brighter and decorated for courtship. Most mammals do not have good color vision, but monkeys do, and in some species males have special colors that help their courtship of females. In many mammals, from dogs to giraffes, males detect females by scent. They also use scent to judge the readiness of females to mate.

In frogs the males attract females by sounds, which can be croaks, peeps, barks, or whistles depending on species. Some can be heard from hundreds of meters away. The males take up residence at a breeding site such as a pond and start calling. Females have ears tuned-in to their own species, and virtually deaf to others, so that they home in on the right males. They are seized with no further ado.

Some courtship behavior is rough. Male tortoises butt other tortoises. Males butt back; females move off. But if a male is persistent, a female may eventually relent and mate.

1 Mandrill

2 European tree frog

The male mandrill (1) is much bigger than the female. He has a longer face with much brighter colors. A courting male European tree frog (2) calls with his throat sac expanded to attract a female. The male three-spined stickleback (3) is bright red and greenish blue in the breeding season. He dances to attract the plumper but duller female into his nest to lay her eggs. In courting, the male two-lined salamander (4) cuts the female's back with his sharp teeth.

4 Two-lined salamander

3 Three-spined stickleback

Courtship in birds

In birds such as ducks and pheasants the males are brightly colored and use their colors in courtship displays. Females are dull colored. In other birds males are dull out of the breeding season, but develop brilliant plumage for mating. Parts of the body that are used in display may be emphasized. Cranes dance and bow with outspread wings in courtship, and their heads have colored markings or crests, their wings colored patches.

In those birds in which both sexes are of similar colors, the birds may identify sex by behavior. Often the first part of courtship is the same as threat behavior. Another male will threaten back. Birds that do not are females. Part of the reason for courtship is that it helps a pair of birds become comfortable with each other, without the aggression that might occur between strangers. Caresses may be part of the process. Pigeons bill and coo, touching and getting rid of any fear. Many bird pairs will groom one another, particularly around the head and neck. Pigeons, lovebirds, and zebra finches show this behavior.

Sometimes during courtship a bird will be torn between wanting to approach a possible mate, and fleeing from the stranger. In these circumstances it may do neither, instead doing something that seems irrelevant, such as scratching. In some birds such as mallards this "displacement" activity has been taken into the courtship ritual.

Another ploy used by some birds is to present their potential mate with a gift of food. A male kingfisher presents a fish to the female he is courting. He even presents it to her the right way around for her to swallow it.

Birds such as peacocks and bowerbirds have bizarre courtship displays. However complex or simple the display of a species, a female does not mate until it is complete. In doves, her eggs will not develop until she is courted.

1 Mallard

2 Black-headed gull

3 Great crested grebe

4 Common tern

5 Sage grouse

In the mallard (1), the male is brightly colored for display. The female, who has to look after the eggs and young, is a camouflage color. Here, a male is showing part of his courtship routine, consisting of ritualized preening movements. A pair of black-headed gulls (2) face away from each other. In this species, head-on showing of the dark face patch is an aggressive signal. So this pair, beginning to court and get to know one another, are taking great care not to show their faces. The great crested grebe (3) has an intricate courtship ritual. Some of the successive stages are shown here, ending with the presentation of a beakful of waterweed as the pair rise out of the water. Both sexes take care of the young, diving for food or letting the chicks ride on their backs. The common tern (4) may pair for life. Here the female sits on the nest site they have chosen, while the male courts her by presenting her with a small fish. In the sage grouse (5), the male is bigger than the female, and is equipped to give a magnificent display. He does this on a mating ground that is used year after year. Other males display here too, and a female chooses a mate from those available. Each male may mate with a number of females; but he does not help rear the brood.

Leader of the pack

◀ Many baboon groups are based upon the harem, with a dominant male having access to a group of females. Young or weak males may be unable to breed. The females, as well as the males, may have an elaborate ranking system. The higher-ranking animals take precedence for tasty food morsels or favored sleeping places.

▶ Life in a herd of Thomson's gazelles. These gazelles live mainly in East Africa, where they are the most common species of gazelle. Two males struggle in a battle for supremacy (right). These animals fight for control of an area of ground that contains a number of females. Two gazelles contesting a territory start by threatening one another with chins raised. Then they put heads down, lock horns, and begin pushing. The weaker gives up and moves away, pursued by the winner; he is not usually injured. The victor is able to mate with any of the females that enter the territory (below left). A strong male will be able to mate with many females and will father many young. A newborn Thomson's gazelle (below right) is left hidden in the grass by its mother, except when she returns to feed it. After a few weeks it can run with the herd.

In many mammals there is competition to find mates and breed. Males fight for females directly, or they may fight for possession of a piece of ground, or "territory," of their own. The territory may be important because it contains females. Some sea lions and fur seals fight over a patch of beach. The winner has the females in it as his harem. Alternatively the territory may be an area that gives enough room and food for a pair to live and bring up young. A gibbon family defends a territory of this sort.

Even within a herd or pack it is rare for all animals to be equal. Usually there is one boss, often the biggest and strongest, and all the rest defer to him. There may be one animal that gives way to the boss, but can lord it over the rest, and so on down to the lowliest. This kind of social organization is sometimes called a "pecking order." It is a useful arrangement because the pack spends little time squabbling. It is already decided who gets the choice of mates or the best food. The leader of the pack does best. In a wolf pack the result is that neither junior males nor junior females are likely to breed.

The ranking may be created by fighting. But there is no advantage in getting hurt, so many rankings are created and kept going by other means. In some species, this involves intimidation. Baboon or hippopotamus males yawn at one another, to show their huge canine teeth. If fights occur, they may be ritual trials of strength, as in deer and antelope, rather than attempts to kill rivals. Often animals have a way of signaling that they are giving in, rather than having to carry on fighting and risk serious injury. The loser in a dog fight presents his throat to a rival or rolls over. This signals that he accepts the victor as boss.

Caring parents

Most mammals and birds invest an enormous amount of energy in producing and caring for their young. This pays off in giving the babies a good start in life and increasing the chances of their survival. Many small birds feed their young many times an hour through all the hours of daylight. Luckily this phase lasts, for most, only two or three weeks. But for some larger birds the job lasts longer. Golden eagles are fed by the parents for about three months before they leave the nest. Albatrosses feed their young for up to one year, returning from long flights at sea to visit the chick.

In most mammals all the work falls on the mother. She is the only one that can provide milk. A mother mouse or vole may need double her usual food intake when feeding babies in the nest. Among the minority of mammals where both sexes help with young are foxes, where the father may bring food to his family. Some mammal mothers are constantly with their young, feeding them, cleaning them, and keeping them warm. Others may leave them and just return for brief periods to feed them, as do rabbits, and deer when fawns are young.

Many parents are very protective of young, and will fight off enemies that they would not tackle on their own account. Blackbirds will dive-bomb cats that come too close to their nest. Wildebeests will try to chase away hyenas that come after their calves. Some birds have special distraction displays that they use to lead predators away from their eggs or young. Plovers shuffle along the ground with one wing held out as though it was broken. The predator goes after the parent, which can escape if it wants, and misses the young.

Mammals and birds that grow up fast and make limited demands on their parents may be produced in relatively big broods. Those with a long period of growing and dependence take more looking after. In such species, a large family would be impossible to rear. Elephants, monkeys, and apes normally have just one baby at a time. Whales and seals, whose babies have to be able to swim very quickly in order to survive, also have single babies. The young are provided with large amounts of very thick fatty milk so they put on weight very quickly.

1

2

◄ (Far left) A long-tailed tit feeds a brood of fledglings sitting on top of a nest. Long-tailed tits take two weeks to make their nests, which are domed with a hole in the side, using hair, moss, and cobwebs and a lining of feathers. They lay up to 12 eggs, which both parents incubate. This takes two weeks. Then the parents work hard for another two weeks, fetching food for the young in the nest. A few days later their work is done. Long-tailed tits often live in pine woodlands. Their numbers are reduced during hard winters.

▲ In ostriches (1) a number of females may lay in one nest, but the male and his main female do the work of incubation and shepherding the young. Later, several families of young may join together under the supervision of a guardian adult. This kind of arrangement is of benefit to all.

In cotton-top tamarins (2) the family is a strong unit. The male, and even older brothers and sisters, often carry the babies and help to groom them. Tamarins are related to marmosets, and live mainly in trees in South America.

Animal communication

Nearly all animals need some way of communicating with others. They may want to signal to others that they are friendly, or to show that they are ready to mate. They may need to signal aggression. They may just want to communicate their presence to other animals, and identify what sort of animal they are. In order to communicate, animals use many kinds of signals. Humans see very well, so visual signals are an easy form of communication for us to understand. Some animals have poor eyes, but good noses or ears. These may use scent and sound signals. Our senses may not be acute enough to smell or hear these signals. This can make it more difficult for us to understand them.

SPOT FACTS

• The sensitive antennae of some male moths can detect a female's scent from many miles away.

• Some fish produce an "alarm substance" if attacked. This alerts others in the school.

• A whale may be able to hear sounds made by another whale many miles away.

• Chimpanzees cannot speak like humans, but some have learned 200 "words" of sign language.

• A dog may have 220 million scent-sensitive cells in its nose. A human has 5 million.

• Woodpeckers pound on trees not only to dig out their nesting holes, but also to communicate with one another by "drumming."

• A dolphin emits up to 300 clicks and whistles per second. The clicks are used for echolocation, but the whistles are used to communicate alarm or excitement.

• Some bottom-dwelling marine fishes communicate with other fish by emitting sounds from their swimbladder.

• Male sea elephants will fight to the death over a group of breeding females.

• The spray of a skunk has such an offensive smell that it repels most predators before they have done any damage.

• Ants have a complex scent code and can guide forager ants to a food source. Assassin bugs can imitate this code and lay false trails, and so eat the ants that are following the trail.

• Black bears from one territory keep their distance from each other mainly by smelling marked trees around their area.

• A young chimpanzee is able to make at least 32 different sounds.

• Howler monkeys roar together morning and night, creating the loudest noise made by any animals.

• One of the most sophisticated forms of invertebrate communication is the "dance" of the honeybee. Using an intricate routine, a field bee can communicate both the distance to, and direction of, a new source of nectar.

• Elephants make rumbling sounds in their nasal passages that are below the range of human hearing. It is believed that they use them to communicate with each other over long distances.

• In troops of capuchin monkeys, it is often the leading male's duty to break up fights between troop members.

• Male deer, when fighting, occasionally lock antlers and die of exhaustion and starvation because they cannot unlock themselves.

• Deer mark their home ground with scent produced by a facial gland in front of each eye.

Smell

▼ Ring-tailed lemurs (1) "fight" by wiping scent from a gland on their forearms onto their tails, then waving the tail at the opposition. Skunks (2) have powerful stink glands to ward off enemies. Ants (3) lay odor trails that can be followed by others, leading to and from good sources of food. Dogs and foxes (4) use scents in urine to mark their territories and pass on other information. Rhinoceroses (5) also use scent to mark territory, leaving piles of dung at the borders of their area. Scent glands may be used to mark other animals as well as defining territories. The Tasmanian devil (6) has glands near the anus. The hyena (7) has them on the chin, legs, and soles of the feet. By sniffing each other, animals such as bighorn sheep (8) can tell a great deal about each other.

Many animals, from insects to fish, use smell to convey messages. For mammals, which usually have poor eyesight, smell is of great importance. Many have special glands in the skin to produce smelly secretions. But with our poor sense of smell we cannot always detect them. When a cat greets you it rubs the side of its face against you. There is a scent gland on its cheek. A rabbit has a gland beneath its chin which can leave a scent mark. Shrews have scent glands on their flanks. Other mammals have them on their backs, between toes, or just in front of the eye, as in many antelope.

Some mammals add scent to their urine or dung. Often their behavior helps make the marker they leave behind more obvious. A male dog cocks its leg to place its scented urine as high as possible. A male hippopotamus wags his tail to spread his dung over a large area of river bank. Scents may be used to proclaim ownership of an area. In some species the animals in a group mark one another so there is a "group scent." In other cases the scent seems to leave a message about the individual that made it. How old or large it is, whether it is ready to breed, even what mood it was in, may all be read by others of the species.

Uses of scent

Sound

Some sounds carry a long way. This makes sound a good way of making contact with another animal. The message sent may mean "keep away," or it may be a call to attract a mate. Some bird songs do both.

It is usually male birds that sing. They are sending a message to other males to keep out of their territory. The singing spaces out the birds so that they do not actually come into contact and fight. Among mammals, howler monkeys and gibbons have a song that carries several kilometers and acts in a similar way.

There are many other messages sent by sound. Some monkeys have alarm calls that alert others of the species. So do marmots. Rabbits drum with their feet to give an alarm. The alarm calls of many birds are quite similar, and can put other species as well as their own on guard. In some kinds of bird each individual has a song with slightly different characteristics so that it can be recognized. Other animals have calls that tell their group they have found food. In chimpanzees there are different calls for different foods.

Whale song

The humpback whale has a complex song (inset) that can last half an hour. Each individual has its own variation.

Sound and hearing are to whales what vision and smell are to most land animals. At least two kinds of sound are produced: rapid echolocation clicks and vocalizations. Echolocation works like a kind of *sonar*. The clicks can be felt by nearby swimmers, and the whale receives the echo and builds up a picture of the local seascape, including information on size, density, and distance.

The humpback whale's song may well be an intricate language, but our understanding of it is in its infancy. It certainly carries for long distances and may well be used to coordinate breeding. Cooperative hunting among killer whales occurs to a degree that would be very difficult without sophisticated communication. Although whales communicate in this way, the 9-kg (20-lb.) brain of the sperm whale is, compared with body mass, proportionally smaller than a human's.

◄ During the rutting season, the red deer stag fiercely defends his harem of female deer against the approach of other male deer. His loud roar to proclaim his presence and ward off rivals carries a long distance.

▶ The male European robin's song, sung from a tall tree at the boundary of his territory, is a warning to other males to keep away. Intruders are fought off. European robins sing nearly all year round. The robin may migrate during the winter months, or may remain resident according to the severity of the winter.

▶ (Far right) The olive baboon has several different alarm barks. These warn others whether the danger is a leopard, snake, or bird of prey.

69

Expressions and gestures

In animals with good eyesight, "conversation" can be carried out by means of using the face and body. We do this ourselves, often without realizing it. Our frowns or smiles tell other people how we are feeling. We may shrug our shoulders, or look bowed down if dejected. Animals, too, have a whole range of body stances and gestures that can be read by others of their own kind.

A confident wolf stands tall, with tail up. One that has just lost a fight will slink away with its tail between its legs. In cats a female may take up a particular stance as an invitation to mate. In chimpanzees reaching out to touch another gently is a sign of friendliness. Flicking a fore-arm at another individual chimpanzee may show irritation with it. There are many different mammals that use "body language." Birds, too, use the position of the body, head, and tail to communicate.

Fewer animals use facial expressions. Not many have the necessary facial muscles. Some carnivores, though, such as dogs and cats, have mobile faces. So do monkeys and apes. Some of the expressions apes use are rather like ours, some are different. In monkeys, grins, pouts, and smacking of lips differ from faces we might make but they are important for communication. Some monkeys have colored eyelids that emphasize their expressions.

Wolf faces

1   2

3   4

5   6

▲ Facial expressions in wolves. Exactly similar ones can be seen in many domestic dogs. A friendly face (1), with ears in mid-position. A submissive expression (2), seen in wolves that give in to a rival, or in a dog that has been reprimanded. A playful face (3), with alert look and ears cocked. An aggressive face (4), in an animal about to attack another. The ears are up. A very defensive face (5), baring the teeth in readiness but with ears down. An aggressively defensive face (6), with lowered ears, wrinkled nose, and lips drawn back into a snarl.

▼ Expressions in cats. A neutral stance (1), with no particular emotion displayed. A friendly cat (2), with head and tail high and ears pricked up. An extremely defensive cat (3), threatening retaliation with claws extended, ears flat, and eye pupils wide. Crouching in a defensive posture (4), with ears down. A cat on guard, showing the defensive threat (5), given to other species, with fur on end and back raised. Threatening attack (6), with legs stiff, tail curved toward opponent, hair partly raised, ears back, and narrowed pupils.

Body language in cats

1 Neutral   2 Friendly   3 Extreme defensive

Macaque language

2

▲ A fight between two silver-backed jackals is about to ensue. The one on the right is aggressive, the other has a defensive face.

◄ Monkey faces. A Barbary ape, a type of macaque, "lip-smacks" at a baby (1). This is a friendly gesture in which the monkey opens and closes its lips several times in succession. The same expression is used by most of the macaque species. It is also used as a greeting between adults, or as an invitation to groom. A moor macaque (2) opens its mouth in a threat, displaying its large canine teeth. Macaques are among the most intelligent of monkeys. They live in Asia, North Africa, and Gibraltar.

6 Offensive threat

5 Defensive threat

4 Defensive

71

Fighting

▲ Fighting deer (muntjacs) slash at each other with their tusks.

▲ Giraffe fights involve prolonged contests of neck wrestling. Each animal tries to use its neck as a hammer or lever to unbalance or knock the other over.

▼ Adult male hippos bite at each other with their huge tusks.

In spite of all the measures they take to space themselves apart and avoid conflict, there are occasions when animals fight. Fighting is probably most common in the breeding season when males come to blows over females. Even then, animals will try to win by bluffing.

Animals that fight are often provided with some protection. Fur seals and baboons have capes of fur. Hippopotamuses and elephant seals have tough skins and padding round their necks, to protect them from sharp teeth.

Many animals appear to have dangerous weapons, like the huge antlers of some deer and the horns of many male antelope. Often these fit together without slipping in the version of wrestling the animals adopt.

Most mammals have ritualized fights in which the kinds of moves made are known to both opponents. Once one begins winning, the other retreats fast. Even so, the constant battles for supremacy may take a huge toll on the health of, for example, a top red deer stag.

▲ Bison and their relatives take part in pushing matches and try to force their opponents backward. Eventually one gives in.

▶ White-tailed deer lock their antlers together and wrestle until one deer is unbalanced.

◄ Many small antelopes such as klipspringers have short, sharp horns used for jabbing opponents.

▲ Although tusks are fearsome weapons, most fights between male elephants are trials of strength in which the contestants push backward and forward, rather than serious attempts to wound.

◄ Male ibex sometimes wrestle and charge one another head-on. Their horns and solid skull can take such knocks. The ridges on the horns stop them from sliding when they clash.

▼ Reedbuck try to twist each other's head to the ground as they fight. The curving horn tips stop them from slipping past each other.

The rhythm of life

Night changes to day and back. Away from the tropics there is a yearly rhythm of seasons. Animals have their own life rhythms that fit in with the rhythms of nature. Many alternate sleep and waking in time with day and night or vice versa. Peaks of body temperature and activity occur at about the same time each day. Many animals have an accurate internal "clock." A dog may appear for its meal at the same time each day. To cope with changes in the seasons, some animals go on long migrations. Others avoid bad conditions by hibernation. Breeding is usually timed to make the most of good food supplies and long days in summer.

SPOT FACTS

• Salmon spend up to 10 years at sea, and then return to the river where they hatched.

• The Arctic tern migrates from the Arctic to the Antarctic and back, a distance of 35,000 km (22,000 mi.) each year.

• The ruby-throated hummingbird, weighing just 3.5 g (0.125 oz.), migrates 800 km (500 mi.) nonstop over the Gulf of Mexico.

• Honeybees are the only bee species that do not hibernate during the winter months.

• A bat may take only half an hour changing from its hibernating to its active state.

• Every year, 5,000 million or more songbirds, waders, and waterfowl make the 320-km (200-mi.) journey southward from Europe to the grasslands, forests, and lakes of Africa. Many cross via the Straits of Gibraltar or the Isthmus of Suez.

• A Manx shearwater taken from its burrow in southwest Wales to Boston, Massachusetts, was back on its nest 12.5 days after being set free in America. Its Atlantic crossing beat the mail carrying details of its release by 10 hours.

• The African lungfish often becomes dormant in a burrow during a prolonged dry period; this may last as long as four years.

• Many bats native to North America spend the winter in Bermuda.

• During its lifespan the monarch butterfly migrates twice between Mexico and the northern United States, a distance of up to 2,900 km (1,800 mi.).

• The green turtle migrates from the coast of Brazil to breed on Ascension Island, 3,340 km (1,400 mi.) away in the Atlantic.

• Some species of grasshopper undergo seasonal color changes, being green at some times and red or brown at others.

• Fur seals are native to many Arctic seas, but at breeding time they all congregate at a rookery on the Pribilof Islands in the Bering Sea.

• Birds probably navigate with a combination of dead reckoning and observation of the Sun and stars. In experiments, bird navigation in cloudy conditions is very erratic.

• The albatross is an extremely fast migratory bird. It can cover some 500 km (300 mi.) a day for 10 days at a stretch.

• The doldrums – the equatorial region characterized by its dead calm seas and very light breezes – are an effective migratory barrier. Owing to the lack of winds, vast amounts of energy are needed to fly across them.

• The woodchuck may spend up to two-thirds of its life in hibernation.

• Tunny fish migrate great distances to spawning and feeding grounds; a fish tagged off California was caught off Japan 10 months later.

Winter and summer

In the tropics it may be warm all year, but animals may have to cope with changes between wet and dry hot seasons. But these changes are usually much less significant than those in cooler regions where there is a big difference between a warm summer and a cold winter. In summer plants grow and there may be plenty of food. In winter there is less to eat.

Small mammals are particularly at risk, because they have a large surface compared with their body size. They lose heat fast, and can end up needing more food than they can find. Some solve the problem by hibernation. At the onset of winter they find a protected retreat and stop activity. They allow their body temperature to drop, almost to the temperature of the surroundings. Their breathing and heart rate slow down. Provided they do not freeze, they can survive the winter in a state of suspended animation. In the spring they warm themselves up again and resume activity.

Animals from hedgehogs to marmots hibernate. Most hibernators put on a lot of fat before their winter sleep. Most is used up, but a little of a special type, called brown fat, is needed for the animal to warm itself again. The energy saving in hibernation is enormous. Perhaps 70 times more energy would be needed to stay active, with no guarantee of survival.

The smaller bats are unusual in that not only do they hibernate over winter, they also "hibernate" each day when they sleep. Hummingbirds do the same thing. It is a way tiny animals can cut down on their energy needs. Large animals generally do not hibernate, but bears may spend the worst part of the winter sleeping in a den with a body temperature a little below normal, to conserve energy.

▲ A Sumatran tiger cools off in a stream. Tigers dislike the great heat of the tropical hot season, and spend much of the day resting or bathing to keep cool.

◄ A dormouse sleeps the winter away. In the fall it makes itself a nest in a frost-free place. In cold climates it may sleep for nine months of the year, emerging only in summer.

Bird migration

Many birds make seasonal migrations. How far they go may depend on the type of food they eat. Seed-eating birds just need to avoid the worst of the weather. Seeds can still be found in the winter. So a seed eater may migrate just a short distance. For an insect-eating bird, though, there may be nothing to eat in the far north in the winter. It needs to fly hundreds or thousands of kilometers south to ensure a supply of insect food. It returns north for a summer in which there will be plenty of insects. There will be long daylight hours in which to catch them and to feed their young. It is worthwhile for a swallow to migrate all the way from northern Europe to southern Africa and back. It pays off in increased survival and numbers of young reared. Northern nectar feeders such as the rufous hummingbird from Alaska also need to migrate. This tiny animal covers 7,000 km (over 4,000 mi.) on its annual round trip.

Some of the distances covered by birds on migration are huge. Willow warblers can travel 12,000 km (7,500 mi.) twice a year. The longest migration route is that of the Arctic tern, which

travels from the Arctic to the Antarctic and back, so that it lives nearly its whole life in daylight. But many other seabirds also make long trips across the oceans. Most migrating land birds avoid sea crossings as much as possible. This is why great congregations of migratory birds can be seen crossing narrow straits such as that between Gibraltar and Africa. A small bird like a warbler can put on enough fat to double its weight before migration. It may then have enough fuel to fly for up to three days without rest.

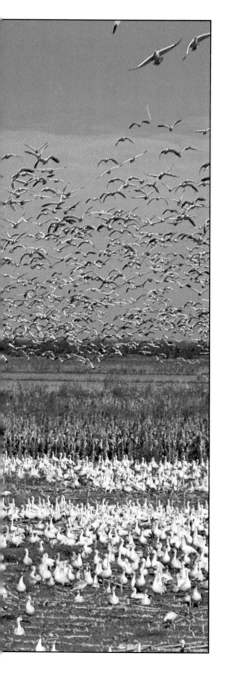

Shearwater migrations

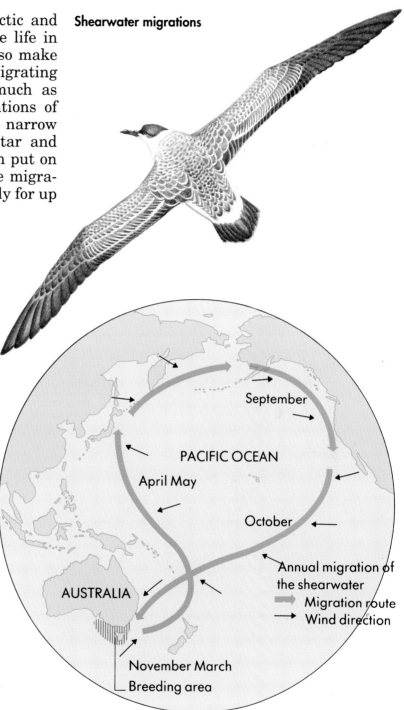

▲ (Top) The great shearwater breeds on the Falkland Islands, then migrates to the north Atlantic and back. (Above) The short-tailed shearwater breeds around south-eastern Australia, then circles virtually the whole Pacific Ocean before returning for the next breeding season.

◀ Huge flocks of snow geese migrate north to make use of the brief Arctic summer for breeding, before returning south to less harsh surroundings for the winter.

Other animal migrations

Many fewer mammals than birds migrate, but some species do make spectacular seasonal movements. Hundreds of thousands of wildebeests migrate across parts of East Africa. Zebras also move over the plains with the seasons. Before farms and other obstacles were in the way, the migration of springbok across southern Africa was one of the biggest animal migrations, with millions taking part. The herds would be passing one spot for days. One of the longest present-day mammal migrations takes place in Canada. Huge herds of caribou move up into the Arctic in the summer to eat the summer growth of plants on the tundra. In the autumn they move back south to a more sheltered winter range. Winter and summer pastures can be 800 km (500 mi.) apart.

Several kinds of marine animal migrate.

Some whales spend the summer in good feeding areas in cold seas. Then they migrate closer to the Equator for the birth of calves and breeding. Some sperm whales migrate 8,000 km (5,000 mi.) in each direction. Salmon make long migrations to breed. An individual salmon will probably go each way once in a lifetime: down river and out to sea when young, and back up the same river when mature to spawn and die.

Some sea turtles travel long distances each year. Green turtles feed off the coast of Brazil, and breed on Ascension Island in the middle of the South-Atlantic Ocean, over 2,000 km (1,200 mi.) away. The turtles migrate between the two areas, but the route they take is not known. If they went direct, they would be traveling in one direction against ocean currents moving faster than they can swim.

◀▼ Elvers (young eels) ready to swim up a river. European eels hatch from eggs in the Sargasso Sea. The leaf-shaped larvae drift in the currents toward Europe, taking four years for the journey. Before arriving they become eel-shaped, but transparent, elvers. They go up the rivers and grow for several years before returning as adults.

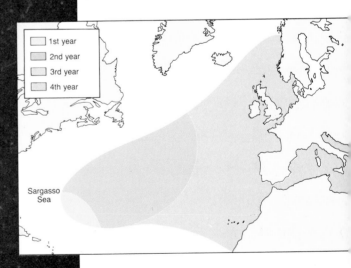

	1st year
	2nd year
	3rd year
	4th year

Sargasso Sea

▶ Wildebeests cross a river on a migration across the Serengeti Plains. Obstacles such as this do not stop the herd, although if the rivers are flooding many animals may get washed away. Wildebeests make a circuit of the plains each year, following the growth of new grasses, which they prefer. Calves are born during the time herds are on the march, and have to be able to walk immediately.

Living together

For many animals there is an advantage in living together. It may make life safer. It may make it easier to find food or a mate. Some animals cooperate in the care of young. In some groups, different jobs can be farmed out to different individuals. The ultimate in specialization and cooperation is found in some insect societies, where individuals could not survive alone. Some animals live together in another sense. They team up with an animal of a completely different species.

SPOT FACTS

• Some African weaverbirds, for protection, build their nests directly above the nests of fierce wasps.

• A nest of honeybees may contain as many as 50,000 individuals.

• An elephant herd is usually led by the oldest female.

• Social birds, such as flamingos and zebra finches, breed well only when in flocks.

• Red-billed queleas may live in flocks of a million.

• Many bugs secrete a sweet-tasting fluid called honeydew. They are often herded, protected, and "milked" for the honeydew by certain species of ants.

• Termites feed mainly on wood, which they can only digest because of the protozoans that live in their intestines.

• Not only does the honey guide bird have a useful relationship with the honey badger, it also gets its unique ability to digest wax from helpful micro-organisms in its gut.

• Meercats come out of their burrows in large crowds, but the outermost ones are always on the lookout for danger. Lookout duty is on a rota system.

• The adults in a musk ox herd, when threatened, form an outward-facing circle with their young in the middle.

• A termite colony may number as many as a million insects. It may have complex structures, including a managed airflow system.

• Some species of dolphin make up groups with tens of thousands of members.

• A beaver dam more than 300 m (1,000 ft.) long was found in Rocky Mountain National Park, Colorado. This would have needed the cooperation of several beaver families.

• The Crocodile bird is so called because the Greek historian Herodotus (484?–425 BC) reported a bird that entered the open jaws of crocodiles to pick bits of food, including leeches, from their gums. This phenomenon has not been widely observed over the centuries.

• Emperor penguins, breeding in some of the most inhospitable places on Earth, take turns to stand on the windward side of the rookery, protecting the others from temperatures as low as -62°C (-80°F).

• A honeybee community consists of three different forms: the queen (female), the drone (male), and the worker (sterile female).

• Male lion cubs are expelled from the pride when they are about three and a half years old.

• In the majority of gregarious mammals, males only join herds for the mating season.

• Small Duiker antelopes, when frightened, dive through the undergrowth and then stand on their hind legs to look around.

• While goats are highly social animals, there is often an old buck (male) that lives alone and acts as an outpost or sentinel on the edge of each herd.

• In Mongolia, herds of wild donkeys may consist of as many as a thousand individuals.

Insect societies

Social insects include termites, the ants, and some of the bees and wasps. Huge numbers may live in the same nest. The whole of their lives is devoted to the well-being of the group rather than the individual.

In termites there is both a king and queen. The queen is comparatively huge, with a swollen abdomen full of eggs. This breeding pair live in a chamber at the center of the nest, surrounded by workers. The workers never mature and cannot breed. A soldier caste with large heads has the job of protecting the nest. Some species bite hard. Soldiers of other species shoot sticky poisons from the front of the head. Termites feed on wood and decaying plants. They can do a great deal of damage to wooden houses. Some species have millions in a nest.

In a honeybee hive there is a single egg-laying female, the queen. The male bee dies after mating. The other females are workers and unable to breed. They are kept busy cleaning the nest and tending the eggs and larvae. They also make wax for the hexagonal cells in which the eggs are laid, larvae reared, and honey supplies stored. Workers forage for nectar and pollen. Some act as guards, stopping strangers from entering the hive. The bees "talk" by touching antennae. A bee returning from a good source of food does a special dance. This shows others where the food is, and how far away. The queen makes a chemical which stops other females from developing. If she dies, or the hive becomes too big, workers start rearing new queens.

Ants are highly successful social insects. Various species run aphid "farms," cultivate fungi as food, forage in huge armies, or make slaves of other ants. Ant nests may be found in many different kinds of places, including under the ground or in treetops.

◀ Weaver ants pulling two leaves together. The larvae secrete silk, and are passed backward and forward to glue the leaves together.

▼ Brood cells in the hive of a honeybee. The larvae are tended by workers. At various stages in their lives workers are baby-sitters, nest guards, and foragers.

Groups and troops

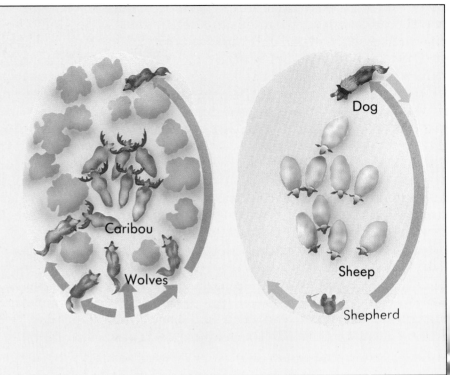

Hunting sheep

When a wolf pack hunts there may be a high degree of cooperation between animals. One may circle around in front of quarry such as caribou, while others drive them forward to the ambush. Another hunting method is to try to split off a young or crippled animal from the rest of the herd.

When a sheepdog is set to work, it uses just the same techniques as its wolf ancestors. Other sheepdogs, or the shepherd, may play the part of the pack, while a single dog is sent in front to hold the sheep in place or drive them back. Separating one sheep from a flock is also part of a sheepdog's job. The ideal sheepdog is a good "hunter" and has its ancestors' instincts, but is gentle or well trained enough to avoid actual attacks on sheep.

Caribou

Wolves

Dog

Sheep

Shepherd

Just good friends

▼ Three savanna baboons get acquainted. The adult male on the right is the "special friend" of the female in the middle. He stays close to her and grooms her. When she is ready to breed he stands a good chance of mating with her. Another male, on the left, is trying to join the troop. To do this he is trying to form a friendship with the female. If she was by herself she might be forced to accept him. But she is unlikely to do so while she has the support and protection of the first male.

Some animal groups arise simply because they are all there doing the same thing at the same time. Hundreds of sea gulls may cluster on a garbage dump or behind a plow, but they are not permanently together. There are many kinds of animals, however, that are truly social in that they try always to live in permanent groups. Some animals breed successfully only if they are part of a group.

In the simplest groups, the individuals may not know one another. In a school of fish one may not know the next, but they like to keep together. In a big flock of finches many may be strangers, but the flock is still useful to the birds. The same goes for big herds of antelope.

But many groups, particularly among mammals, are much more permanent and structured. A herd of elephants can continue for many years with much the same membership. The herd leader is the mother or grandmother of many of the animals in it. As males mature, they leave the herd, but females remain even when they are adult. So a herd builds up of sisters and cousins that are well known to one another. Females act as "aunts," helping to look after the offspring of their relatives.

Social groupings

A baboon troop may contain 30 or 40 animals that are known to each other. Each will know its place in the troop, with both males and females having some sort of ranking order. The main adult male will probably be the father of many of the younger troop members. When the troop moves he stays in the middle, surrounded by females and young. Lesser males walk on the edges of the troop with the job of guarding the rest of the baboons.

In most social mammals the role of individuals changes as they grow up. But the naked mole rat, which lives underground in groups of about 50, is different. One female produces all the babies. Others act as workers, much like social insects.

▶ The fish in a school are not social animals in the same sense as, say, baboons. They do not have any complicated ranking system or recognize and react to one another as individuals. Nevertheless, they can appear well drilled, hundreds turning almost in unison. They live together because there is safety in numbers.

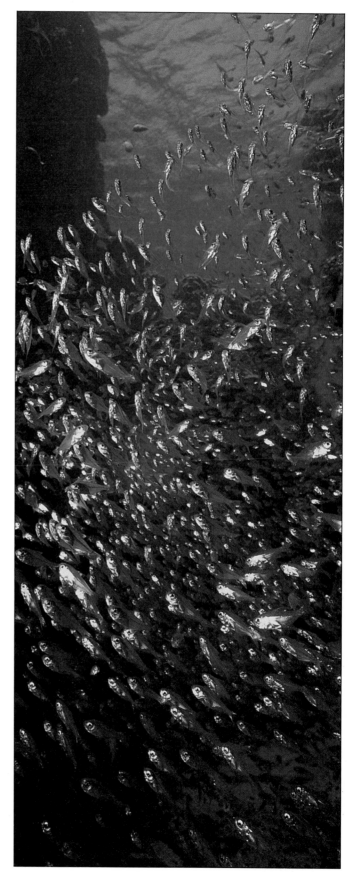

Associations between species

Sometimes, when one species associates with another, only one gains an advantage. When a flea lives on the skin of a dog it is good for the flea but not for the dog. But there are plenty of examples of species that live together with gains for both. Dogs and humans can live together and both benefit. The dog gets free meals in exchange for its work as a guard.

Hermit crabs live in old whelk shells. On the shell may live a sea anemone. Its stinging tentacles help protect the crab. Pieces of food that the crab is tearing apart reach the anemone. Sometimes there is a third partner; segmented worms sometimes also live in old whelk shells.

Another good team is made in warm seas by a goby that shares a burrow on the seabed with a shrimp. The shrimp does most of the work of excavating and keeping the burrow clear. The goby stands guard, just touching the shrimp. If it sees danger, it darts back to the burrow, warning the shrimp, which also dives for cover.

One of the most remarkable partnerships is between the honey guide, an African bird, and the honey badger. The bird likes to feed on bee larvae, but cannot break into nests. If it finds a nest, it may fly to attract the attention of a honey badger, then lead it to the nest. The

Adopted mothers

Young graylag geese follow "mother," in this case Konrad Lorenz, one of the pioneers of the study of behavior in animals. The young of geese, ducks, sheep, and some antelope can run around almost as soon as they hatch or are born. It is important that they form a strong attachment to their mother so that they can follow and stay with her.

They are born with a tendency to follow the first large moving object they see. Normally, of course, this will be their mother. But in captivity it is possible to fool this instinct, if the first animal they see is from another species. So they can be "imprinted" on a scientist, or, as in the case of the ducklings below, on a retriever dog, which seems to enjoy the role.

honey badger has strong claws and a very tough skin that protects it against bee stings. It breaks into the nest and the two animals share the spoils.

In some partnerships only one partner seems to gain anything. The remora is a fish with a sucker it uses to hitch a ride on sharks. It feeds on scraps from the shark's food, but does no service in return. But it does no harm either, and is not a parasite.

▲ An African buffalo grazes with a cattle egret by its side and two oxpeckers on its back. The egret catches insects stirred up by the buffalo. The oxpeckers remove ticks from its back. The birds may alert the buffalo to danger.

▶ A huge sweetlips dwarfs the cleaner wrasse in its mouth. But the wrasse is not dinner. It is performing the service of cleaning the mouth and gills of the bigger fish, taking off parasites that cling there. Big fish, including predators, line up at "cleaning stations" on the coral reef for the cleaning fish's services, and do not attempt to harm them. The big fish hold their mouths and gills open so they are easier to clean. Cleaner fish have clear markings so they can be easily recognized, and not be mistaken for prey.

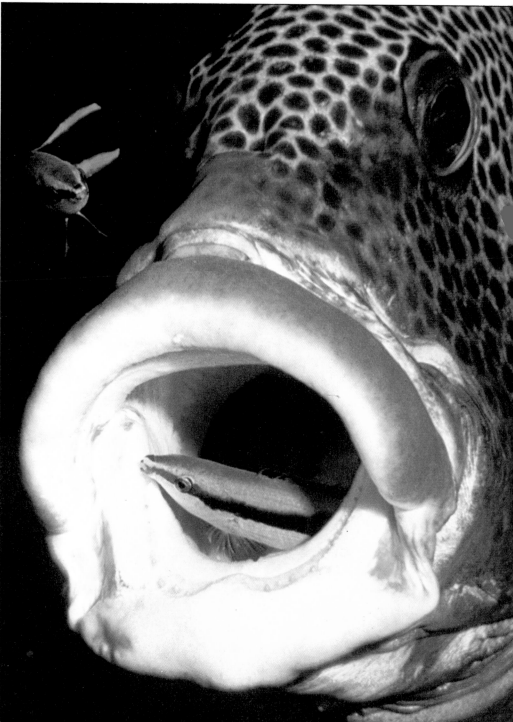

Units of measurement

Units of measurement

This encyclopedia gives measurements in metric units, which are commonly used in science. Approximate equivalents in traditional American units, sometimes called U.S. customary units, are also given in the text, in parentheses.

Some common metric and U.S. units

Here are some equivalents, accurate to parts per million. For many practical purposes rougher equivalents may be adequate, especially when the quantity being converted from one system to the other is known with an accuracy of just one or two digits. Equivalents marked with an asterisk (*) are exact.

Volume
1 cubic centimeter = 0.0610237 cubic inch
1 cubic meter = 35.3147 cubic feet
1 cubic meter = 1.30795 cubic yards
1 cubic kilometer = 0.239913 cubic mile

1 cubic inch = 16.3871 cubic centimeters
1 cubic foot = 0.0283168 cubic meter
1 cubic yard = 0.764555 cubic meter

Liquid measure
1 milliliter = 0.0338140 fluidounce
1 liter = 1.05669 quarts

1 fluidounce = 29.5735 milliliters
1 quart = 0.946353 liter

Mass and weight
1 gram = 0.0352740 ounce
1 kilogram = 2.20462 pounds
1 metric ton (tonne) = 1.10231 tons

1 ounce = 28.3495 grams
1 pound = 0.453592 kilogram
1 short ton = 0.907185 metric ton (tonne)

Length
1 millimeter = 0.0393701 inch
1 centimeter = 0.393701 inch
1 meter = 3.28084 feet
1 meter = 1.09361 yards
1 kilometer = 0.621371 mile

1 inch = 2.54* centimeters
1 foot = 0.3048* meter
1 yard = 0.9144* meter
1 mile = 1.60934 kilometers

Area
1 square centimeter = 0.155000 square inch
1 square meter = 10.7639 square feet
1 square meter = 1.19599 square yards
1 square kilometer = 0.386102 square mile

1 square inch = 6.4516* square centimeters
1 square foot = 0.0929030 square meter
1 square yard = 0.836127 square meter
1 square mile = 2.58999 square kilometers

1 hectare = 2.47105 acres
1 acre = 0.404686 hectare

Temperature conversions

To convert temperatures in degrees Celsius to temperatures in degrees Fahrenheit, or vice versa, use these formulas:

Celsius Temperature = (Fahrenheit Temperature − 32) × 5/9
Fahrenheit Temperature = (Celsius Temperature × 9/5) + 32

Numbers and abbreviations

Numbers

Scientific measurements sometimes involve extremely large numbers. Scientists often express large numbers in a concise "exponential" form using powers of 10. The number one billion, or 1,000,000,000, if written in this form, would be 10^9; three billion, or 3,000,000,000, would be 3×10^9. The "exponent" 9 tells you that there are nine zeros following the 3. More complicated numbers can be written in this way by using decimals; for example, 3.756×10^9 is the same as 3,756,000,000.

Very small numbers – numbers close to zero – can be written in exponential form with a minus sign on the exponent. For example, one-billionth, which is 1/1,000,000,000 or 0.000000001, would be 10^{-9}. Here, the 9 in the exponent –9 tells you that, in the decimal form of the number, the 1 is in the ninth place to the right of the decimal point. Three-billionths, or 3/1,000,000,000, would be 3×10^{-9}; accordingly, 3.756×10^{-9} would mean 0.000000003756 (or 3.756/1,000,000,000).

Here are the American names of some powers of ten, and how they are written in numerals:

1 million (10^6)	1,000,000
1 billion (10^9)	1,000,000,000
1 trillion (10^{12})	1,000,000,000,000
1 quadrillion (10^{15})	1,000,000,000,000,000
1 quintillion (10^{18})	1,000,000,000,000,000,000
1 sextillion (10^{21})	1,000,000,000,000,000,000,000
1 septillion (10^{24})	1,000,000,000,000,000,000,000,000

Principal abbreviations used in the encyclopedia

°C	degrees Celsius		kg	kilogram
cc	cubic centimeter		l	liter
cm	centimeter		lb.	pound
cu.	cubic		m	meter
d	days		mi.	mile
°F	degrees Fahrenheit		ml	milliliter
fl. oz.	fluidounce		mm	millimeter
fps	feet per second		mph	miles per hour
ft.	foot		mps	miles per second
g	gram		mya	millions of years ago
h	hour		N	north
Hz	hertz		oz.	ounce
in.	inch		qt.	quart
K	kelvin (degree temperature)		s	second
			S	south
			sq.	square
			V	volt
			y	year
			yd.	yard

Glossary

abdomen The hind part of the body of an animal such as an insect or crustacean, behind the thorax and containing the sex organs and much of the digestive system. In vertebrates the part of the body containing the stomach and intestines.

adaptation Any feature of a plant or animal that helps it to live in its surroundings.

aggression A hostile kind of behavior usually seen in animals which are in competition with each other for food, territory, or mates.

alarm call A call made by an animal that alerts others of its own or other species to danger. Animals as diverse as birds, deer, and monkeys use calls of this type.

amphibian An animal of the class Amphibia. Most have a larval stage dependent on water and an adult stage that lives on land. Their skins are smooth and lack scales.

ancestor An animal or plant that lived in former times which is thought, through many generations of offspring, to have led to a species living today.

angiosperm A flowering plant.

antennae A pair of appendages on the head of an insect or crustacean. They carry sense organs and may be sensitive to smell, sound, or touch.

antlers Paired structures on the head of deer, made of bone and often branched. They are normally borne just by males and are shed each year.

atmospheric Describes a substance that exists in the air around us.

barb One of the branches coming out of the shaft of a feather.

barbule One of the tiny parts making up the fringe of a barb of a feather. In most flight feathers the barbules can interlock.

bask To sit in the sun, as do many reptiles, in order to raise the body temperature.

brown fat A type of fat found particularly in hibernating mammals. It provides an energy supply to warm up the animal after its winter sleep.

browse To pick leaves and bark from the twigs of bushes and trees, as does a giraffe.

buoyant Able to float or maintain a constant level in the water.

camouflage Colors, patterns, and behavior that help conceal an animal in its surroundings.

canine Like a dog. The canine teeth are in the front corners of the mouth. In carnivores they are long and pointed for killing prey.

carcase The body of a dead animal.

carnivore An animal that feeds on the flesh of other animals.

carpel The female part of a flower. It contains the ovary. There may be many carpels in one flower.

carrion The flesh of a dead animal that has died or has previously been killed by another animal, used as food by a scavenger.

cartilage The soft gristly tissue which is the forerunner of bone in many vertebrates. In sharks the skeleton is made of it.

cartilaginous Composed of or containing cartilage. The cartilaginous fish include the sharks and rays.

caste Those members of a species that have a body, and behavior, that fits them to do a particular job, as for example the soldier caste in termites.

caudal Belonging to the tail. The tail fin of fish is known as the caudal fin.

chloroplast The part of the cell in a plant or green alga that contains the green chlorophyll pigment that traps light energy.

cilia (singular, **cilium**) The tiny hairlike projections from the surface of some cells and single-celled animals. By beating they may move particles across the cell surface, or in single-celled animals they provide a means of locomotion.

cold-blooded Unable to produce heat within the body to maintain a constant body temperature, relying instead on warmth from the surroundings.

colonize To move into a new area.

colony An organism that consists of a number of individuals that live together permanently.

communal Shared in common with other members of the species. Some species have communal burrows; others may have communal mating grounds.

communication Any means of conveying information from one animal to another. Animals may communicate by gestures and other visual signals, by sounds, or by scent.

conifer A plant belonging to the group including pines, firs, and cypresses, where the pollen is borne on male cones and the seeds on female cones. Most conifers have needlelike leaves.

courtship Behavior in animals ready to mate. Usually the male courts the female, making her more willing to mate. Courtship behavior is often ritualized.

cud The ball of food brought up from the stomach into the mouth for a second chewing after partial digestion in the stomach by animals such as cows, giraffes, and camels.

displacement activity Behavior by an animal that seems irrelevant to its situation, as for example scratching itself or pecking at the ground as though for food when it is torn between fighting and fleeing.

display A pattern of behavior, often rather rigid in form, that is performed by an animal and by which it informs another animal of its feelings and intentions. Courtship, greeting, or threatening often involve displays between animals.

disruptive coloration A pattern of colors that break up the outline of an animal's body, making it harder to see.

dominance The ability of an animal to take precedence over another one in disputes over food, mates, or other matters. Dominance may be established by fighting.

dominant Describes an animal that shows dominance over another one.

dormancy A condition in which animals' bodies become cooler and their body systems slow down. The animal appears to sleep. Dormancy saves energy in times of hardship.

embryo A baby animal still within its egg, or, in mammals, in the earliest stages of development within its mother. The earliest stages of development of a plant after fertilization. A flowering plant embryo may rest without further development in the seed for some time.

endosperm Food within a seed which is used up in the seed's development.

eyespot Patch of color on an animal resembling a large eye. It confuses other animals, especially predators.

fertilized Describes a female cell or egg that has been joined by a male cell or sperm that has fused with it.

filament A fine hairlike structure. Some algae form filaments as they grow.

flagella (singular, **flagellum**) Long threads sticking out from a cell that can be moved in a whiplike fashion. They provide a means of moving for some protozoans.

fledging The process of a young bird beginning to fly and leaving the nest.

fledgling A bird that has reached the age when it is beginning to fly. This often coincides with the time it leaves the nest.

flipper A limb which has been modified into a paddle for use in water. Seals and whales have flippers, but these still show the basic arrangement of bones found in land mammals.

fluke A type of flatworm that lives as a parasite inside the body of another animal. Also, one of the tail "fins" of a whale.

forager An animal that forages for food, going out and searching for it. Ants, for example, go out from their nests to look for food.

frond The leaf of a fern, or the flattened body of some types of seaweed.

fruiting body The part of a fungus on which the spores are borne. In certain fungi the fruiting body is called a mushroom.

germinate When a seed or spore starts into growth it is said to germinate.

gestation period The period during which a baby mammal grows inside its mother's body, from fertilization to birth.

gesture A movement of the limbs or body that shows the feelings of an animal, and can convey those feelings to another individual.

gill The breathing organ of an aquatic animal. It contains an area of thin membrane through which gases can easily pass to and from the surrounding water.

gizzard A muscular part of the digestive system of an animal which helps to grind up the food.

gland A group of cells in the body that produce a special substance. Glands may produce substances that work inside the body, or they may release secretions to the outside, as do sweat glands and scent glands.

groom To clean or comb the skin and fur. Particularly used of one animal performing this service for another, as when one baboon grooms another.

habitat The surroundings in which an animal or plant lives, including the other animal and plant life, the physical surroundings, and the climate.

harem A group of females with which a male surrounds himself and with which he mates. Elephant seals, red deer, and baboons are among the animals that have a harem system of mating.

herbivore An animal that eats plants.

hexagonal Having six sides, like the cells in a honeycomb.

hibernate To go into a period of winter sleep, or dormancy. Hedgehogs, dormice, and bats all hibernate.

honeydew A sweet sticky substance secreted by aphids and certain other tiny insects. It is a favorite food of some ants.

horns Paired appendages found on the head of such animals as antelope, giraffes, and oxen. They have a horny sheath with a bony core at the base. They are kept from year to year. In rhinoceroses horns are formed in the midline of the nose and are horny outgrowths of the skin.

imprinted Describes the behavior of a young animal that has learned early in life the characteristics of an object (usually the mother) which it thereafter follows automatically. In experiments it is possible to make newly hatched or newborn animals follow unlikely objects if these are "imprinted" by being the first things they see.

incubate To sit on eggs to keep them warm and protect them.

incubation The period of development of an egg between its laying and its hatching.

instinct Behavior which is inherited rather than learned, so that it is built-in in an animal. Suckling is instinctive behavior of a baby mammal. It must be born with this behavior if it is going to survive.

invertebrate An animal without a backbone.

king The adult reproductive male of a termite colony. The king continues to live with the reproductive female, or queen, for periods of up to several years.

larvae The young stages of fish and amphibians, as well as insects and other animals without a backbone, particularly when they differ greatly in appearance from the adults. In butterflies and moths the larva is a caterpillar.

lateral line A line of sense organs that runs down the side of a fish. It can feel movements in the water.

mantle cavity A space in a mollusk's body connected to the surrounding water. Often the gills are situated there. Some land snails use this cavity as a lung.

marsupial A mammal in which the baby is born in a very undeveloped state. It attaches itself to the mother's teat, often within the protection of a pouch.

membrane A thin skin covering a body, organ, or cell.

migrate To travel in a migration.

migration Long-distance animal movements made on a seasonal basis, as between summer nesting and winter feeding grounds.

mimic (1) To imitate behavior, as when one bird imitates the song of another. (2) An animal that mimics another. (3) An animal that so resembles another species in shape and color that it may be mistaken for the other. Usually it is gaining an advantage by any confusion that may be caused.

mimicry The way that animals of one species gain an advantage by looking like animals of another species.

molt To shed a coat of fur or feathers, or a skin, generally having it replaced with another.

notochord A stiffening rod that runs down the back of certain animals, known as chordates, at some stage in their life history. Among the chordates are animals with backbones. In these, the notochord is replaced by the backbone during development.

ovary The body organ in which female sex cells are produced.

parasite An animal or plant that lives on or inside another organism, living at its expense and providing nothing in return. Fleas and tapeworms, for example, are parasites.

pectoral On the chest. The front pair of fins on a fish are called pectorals.

pelvic On the hips. The rear pair of fins on a fish are called pelvics.

petal One of a ring of flattened, often brightly colored structures around the edge of a flower. Most seem to attract insects.

phylum (plural, **phyla**) The biggest group into which animals are divided in classification. The phylum includes all the animals

with the same basic body plan. For example, all segmented worms are placed in the phylum Annelida.

placental A mammal in which the baby grows inside the mother, attached to her womb by a placenta through which it is nourished. In such a mammal the baby is born relatively well developed.

predator Any animal, such as a lion or an eagle, that feeds by catching and killing another.

prey Any animal that becomes a meal for a predator.

primaries The long flight feathers at the tip of a bird's wing, carried on the skin of the fingers.

pungent Having a sharp, biting effect on the nose or taste buds.

pupil The black center of the eye. It is the gap in the iris that lets light through to the back of the eye.

quarry The target of a hunt. An animal that a hunter is trying to catch.

queen A fertile adult reproductive female of a species of social insect, such as the queen that lays all the eggs in a honeybee hive.

ranking The position that an animal has within a group. A high-ranking animal will be dominant over a low-ranking individual.

ritual A pattern of behavior that is fixed in the way it is performed. Its meaning is recognized by other members of a species, but may not be immediately obvious to an observer.

ritualized Describes behavior that has become a ritual, or part of a ritual. It may in the process have lost its original meaning or purpose. For example, ritual preening has become part of the courtship rituals of some ducks.

ruff A frill of feathers or skin

around the neck of an animal, particularly such a frill that can be extended outward.

rutting season The period of the year when males round up and mate with females. Used particularly of deer and antelope in which the males gather harems and challenge one another.

scavenge To search for dead or decaying animal or plant matter as food.

secondaries The inner, shorter flight feathers on a bird's wing.

secrete To produce and give off a substance in the body, particularly from a specialized gland. Thus, saliva is secreted from a salivary gland.

segmented Describes a body made up of a series of repeated rings, or segments, each rather similar to the next, such as the rings of an earthworm. Insects, crustaceans and spiders are also basically segmented, but in some of these, particularly spiders, the segmentation may be harder to observe.

sepal One of the ring of flower parts, usually green, sometimes colored, outside and below the petals.

shaft The main axis of a feather.

silica A hard mineral, silicon dioxide, that makes up sand and quartz. It is found in sponges and grasses.

social Living in a group together with others of the same species. If an animal is social it is not necessarily in harmony with all members of its group.

species All the animals of the same kind. They have the same structure and can breed together.

spore A stage in a plant's life history adapted for rest or dispersal. It is usually single-celled, with a tough skin, and is produced without sexual reproduction.

spore capsule A hard-walled part of a plant containing spores.

stamen The male part of a flower where pollen is produced.

stance The way that an animal stands.

submissive Giving way to another animal. Animals may have submissive behavior, for example a particular posture, that signals that they will yield to a dominant individual.

suspended animation A state in which an animal is alive but seems totally inactive.

swim bladder (or **air bladder**) A balloonlike organ in a fish's body that is filled with air or gas. In most fish it helps control buoyancy.

symmetrical With exactly similar parts in two or more directions.

taproot A main root of a plant going straight down into the ground.

territory The area in which an animal or group of animals lives and which it defends against intruders of the same species.

testes The body organs in which male sex cells are produced.

thorax The chest region. The middle of the three regions into which insect bodies are divided.

tube feet The small water-filled extensions from the body of a starfish or sea urchin on which the animal walks.

tundra A type of vegetation found in the colder regions of the world toward the poles and near the tops of mountains. It consists of low-growing shrubs, mosses, lichens, and some cushion-shaped flowering plants. Nothing grows tall or has deep roots.

vane The flat part of a feather.

vascular system A system of tubes within a plant or animal. The water and food-carrying tubes within plants, or the blood system of animals.

venom A poisonous substance produced by an animal, particularly one which is injected by biting or stinging.

vertebrate An animal with a backbone. Amphibians, reptiles, fish, birds, and mammals are vertebrates.

warm-blooded Describes an animal that can maintain a constant body temperature in all kinds of surrounding temperatures.

Index

Further reading

Baker, Wendy, and Andrew Haslam. *Insects.* New York: Thomson Learning, 1994.

Carwardine, Mark. *Whales, Dolphins, and Porpoises: The Visual Guide to All the World's Cetaceans.* New York: Dorling Kindersley, 1995.

Catherall, Ed. *Exploring Plants.* Chatham, NJ: Raintree Steck-Vaughn, 1992.

Cavendish, Marshall. *Wildlife of the World.* Tarrytown, NY: Marshall Cavendish, 1993.

Dal Sasso, Cristiano. *Animals: Origin and Evolution.* Chatham, NJ: Raintree Steck-Vaughn, 1994.

Forsyth, Adrian. *Exploring the World of Insects: The Equinox Guide to Insect Behavior.* Buffalo, NY: Firefly Books Ltd., 1992.

Forshaw, Joseph. *Birding.* Alexandria, VA: Time-Life Books, 1995.

Fourie, Denise. *Hawks, Owls and Other Birds of Prey.* Morristown, NJ: Silver Burdett Press, 1995.

Ganeri, Anita. *Insects.* Danbury, CT: Franklin Watts, 1993.

Ganeri, Anita. *Plants.* Danbury, CT: Franklin Watts, 1992.

Greenaway, Theresa. *Trees.* New York: Dorling Kindersley, 1995.

Gutfreund, G. *Vanishing Animal Neighbors.* Danbury, CT: Franklin Watts, 1993.

Halpern, Robert R. *Green Planet Rescue: Saving the Earth's Endangered Plants.* Danbury, CT: Franklin Watts, 1993.

Hunt, Joni P. *Insects.* Morristown, NJ: Silver Burdett Press, 1994.

Illustrated Encyclopedia of Wildlife. Lakeville, CT: Grey Castle Press, 1990.

Kerrod, Robin. *Animal Life.* Tarrytown: Marshall Cavendish, 1993.

Landau, Elaine. *Endangered Plants.* Danbury, CT: Franklin Watts, 1992.

Parker, Steve. *Fish.* New York: Random House Books for Young Readers, 1990.

Patent, Dorothy H. *How Smart Are Animals?* Orlando, FL: Harcourt Brace & Co., 1990.

Pope, Joyce. *Plant Partnerships.* New York: Facts on File, 1991.

Ricciuti, Edward. *Fish.* Boulder, CO: Blackbirch, 1993.

Silverstein, Alvin, et al. *Plants.* New York: Twenty-First Century Books, 1995.

Steele, Philip. *Fish.* Morristown, NJ: Silver Burdett Press, 1991.

Time-Life Books. *Insects and Spiders.* Alexandria, VA: Time-Life Books, 1995.

Picture Credits

b=bottom, t=top, c=center, l=left, r=right.

A Ardea, London. ANT Australasian Nature Transparencies, Victoria, Australia. BCL Bruce Coleman Ltd., Uxbridge, Middlesex. NHPA Natural History Photographic Agency, Ardingly, Sussex. OSF Oxford Scientific Films, Long Hanborough, Oxfordshire. PEP Planet Earth Pictures, London. SPL Science Photo Library, London.

6 Zefa/G.Heilman. 9c Microscopix/Andrew Syred. 9l SPL/Dr. T. Brain & D. Parker. 9r SPL/Dr. Kari Lounatmaa. 10 OSF/Peter Parks. 11 OSF/Barne E. Watts. 12 Premaphotos/K. Preston-Mafham. 13t G. R. Roberts. 13b NGPA/John Shaw. 15l BCL/Hans Reinhard. 15r Iriarte. 16 Biofotos/Heather Angel. 17 SPL/Dr. Eckart Pott. 19 Graham Bateman. 22t PEP/Nancy Sefton. 22b PEP/K. Vaughan. 23l R. W. Van Devender. 23r A/P. Morris. 26 Natural Science Photos/I. Bennett. 27 Jacana/Chaumeton. 29 Anthony Bannister. 30 Agence Nature/Chaumeton/Labat. 31 A/V. Taylor. 37 A/J. P. Ferrero. 41 C. A. Henley. 46 Zefa/Hackenberh. 49 Premaphotos/K. Preston Mafham. 52t PEP. 52b BCL/Jane Burton. 52-53 Anthony Bannister. 54 Michael Fogden. 55t PEP/Alan Colclough. 55bl Michael Fogden. 55br Anthony Bannister. 56 A/Don Hadden. 57t OSF/David Shale. 57bl BCL/Hans Reinhard. 57br Michael Fogden. 62l R. Dunbar. 62-63 PEP. 63bl Jacana. 63br William Ervin/Natural Imagery. 64 Swift Picture Library/M. King. 68 BCL/Hans Reinhard. 69t A/F. Gohier. 69 (inset) R. Payne/American Association for Advancement of Science. 69cb BCL/N. G. Blake. 69br PEP/A. & M. Shah. 71 Nature Photographers. 75lBCL. 75r BCL. 76-7 Leonard Lee Rue. 78 Biofotos/Heather Angel. 79 Survival Anglia/John Pearson. 81l ANT/G. A. Wood. 81r A/John Mason. 83 PEP/P. Scoones. 84-85 Jacana/J. Robert. 84bl H. Kacher. 84br Ann Cumbers. 85 PEP/B. Wood.